Springer Tracts in Mechanical Engineering

Springer Tracts in Mechanical Engineering (STME) publishes the latest developments in Mechanical Engineering - quickly, informally and with high quality. The intent is to cover all the main branches of mechanical engineering, both theoretical and applied, including:

- Engineering Design
- Machinery and Machine Elements
- Mechanical Structures and Stress Analysis
- Automotive Engineering
- Engine Technology
- Aerospace Technology and Astronautics
- Nanotechnology and Microengineering
- Control, Robotics, Mechatronics
- MEMS
- Theoretical and Applied Mechanics
- Dynamical Systems, Control
- Fluids Mechanics
- Engineering Thermodynamics, Heat and Mass Transfer
- Manufacturing
- Precision Engineering, Instrumentation, Measurement
- Materials Engineering
- Tribology and Surface Technology

Within the scope of the series are monographs, professional books or graduate textbooks, edited volumes as well as outstanding PhD theses and books purposely devoted to support education in mechanical engineering at graduate and post-graduate levels.

Indexed by SCOPUS, zbMATH, SCImago.

Please check our Lecture Notes in Mechanical Engineering at https://link.springer.com/bookseries/11236 if you are interested in conference proceedings.

To submit a proposal or for further inquiries, please contact the Springer Editor **in your region**:

Ms. Ella Zhang (China)
Email: ella.zhang@springernature.com
Swati Meherishi (India)
Email: swati.meherishi@springer.com
Dr. Leontina Di Cecco (All other countries)
Email: leontina.dicecco@springer.com

All books published in the series are submitted for consideration in Web of Science.

Liang-Chia Chen · Guo-Wei Wu ·
Sanjeev Kumar Singh · Wei-Hsin Chein

Diffractive Image Microscopy for 3D Imaging

Springer

Liang-Chia Chen
National Taiwan University
Taipei, Taiwan

Guo-Wei Wu
National Taiwan University
Taipei, Taiwan

Sanjeev Kumar Singh
National Taiwan University
Taipei, Taiwan

Wei-Hsin Chein
National Taiwan University
Taipei, Taiwan

ISSN 2195-9862 ISSN 2195-9870 (electronic)
Springer Tracts in Mechanical Engineering
ISBN 978-981-97-7781-5 ISBN 978-981-97-7782-2 (eBook)
https://doi.org/10.1007/978-981-97-7782-2

This Springer imprint is published by the registered company Springer Nature Singapore Pte Ltd.
The registered company address is: 152 Beach Road, #21-01/04 Gateway East, Singapore 189721, Singapore

Preface

Optical profilometry is widely used as its noncontact nature makes it nondestructive. An optical system's design and principle can provide features of point, line, or area scanning, resulting in outstanding flexibility and measurement speed. However, due to its noncontact feature, the measured results cannot always accurately represent the ground truth of the measured object. In particular, some errors are determined by the measurement system itself and depend on the unknown properties of the measured surface (e.g., the surface slope). This error is not easy to remove because the surface profile is unknown before measurement but already contains profile-dependent error after a measurement. To this problem, many calibration strategies have been developed over the years. However, the effectiveness of these proposed methods is limited by many measuring constraints.

In contrast, a new technology called diffractive image microscopic profilometry and its associated measuring microscope have been developed. This technology first provides a focused illumination on the detected surface and then collects the diffractive image of the signal light. A prebuilt diffractive image database combining the image comparison method, the feature comparison method, and the multilayer perceptron has been developed to convert the diffractive image into the surface height and orientation of the measured point. The height, tilting angle, and direction are measured simultaneously and independently, thus avoiding the slope-dependent error.

On top of diffractive image profilometry (DIP) introduced in the first half of the book, another novel technique that incorporates the use of diffractive images for improving the lateral and axial resolution of conventional chromatic confocal microscopes is presented, which is called chromatic confocal diffractive image profilometry (CC-DIP). In CC-DIP, a scalar diffraction parametric model is developed to derive the wavelength-encoded point spread function of a chromatic confocal system. Furthermore, deep learning-assisted deconvolution is applied to the diffractive spectral images obtained from an imaging spectrometer, resulting in improved spatial resolution and reduced signal cross-talks in reconstructed surface profiles.

This book first introduces the fundamental optical principles and provides a comprehensive overview of state-of-the-art optical profilometry techniques.

Following, thorough descriptions of the core principles, practical implementation, and applications of DIP and CC-DIP are given to prepare readers with both theoretical and practical aspects of diffractive image microscopy. With the presented contents in this book, we show that diffractive image microscopy offers excellent potential in pushing the limits of optical profilometry, making it a useful technology for a wide range of research and industrial applications.

Taipei, Taiwan

Liang-Chia Chen
Guo-Wei Wu
Sanjeev Kumar Singh
Wei-Hsin Chein

Contents

Abbreviations

ADAM	Adaptive moment estimation
AFM	Atomic force microscopy
AI	Artificial intelligence
ANN	Artificial neural network
AOI	Automated optical inspection
BADS	Bayesian adaptive direct search
BCE	Binary cross-entropy
BFL	Back focal length
BGA	Ball grid array
BSM	Beam steering mirror
CCD	Charge couple device
CCDIP	Chromatic confocal diffractive image profilometry
CCE	Categorical cross-entropy
CCM	Chromatic confocal microscopy
CCSI	Chromatic confocal spectral interferometry
CMM	Coordinate measuring machines
CMOS	Complementary metal oxide semiconductor
CNN	Convolution neural network
CSI	Coherence scanning interferometry
CTF	Coherent transfer function
DBM	Database building module
DI	Diffractive image
DIP	Diffractive image profilometry
DMD	Digital micromirror device
DOF	Depth of field
DRC	Depth response curve
EF	Extracted features
EFL	Effective focal length
ELU	Exponential linear unit
EM	Electromagnetic
FDB	Squared fractional polynomial image database

FEM	Finite element method
FOV	Field of view
FRRMS	Fitting residual root mean square
FV	Focus variation
FWHM	Full width
GAN	Generative adversarial network
GDB	Gaussian database
GUI	Graphic user interface
GV	Grayscale value
IDB	Image database
INCC	Interpolated NCC
IPS	Ideal point source
ITF	Imaging transfer function
IZNCC	Interpolated ZNCC
LED	Light emitting diode
LMA	Levenberg-Marquardt Algorithm
MAE	Mean absolute error
MCM	Monochromatic confocal microscopy
MEMS	Microelectro mechanical system
ML	Machine learning
MLP	Multiple layer perceptron
MNCC	Maximum NCC-looking method
MSE	Mean square error
MTF	Modulation transfer function
MZNCC	Maximum ZNCC-looking method
NA	Numerical aperture
NCC	Normalized cross-correlation
NFV	Normalized feature vector
NPP	Normalized posture parameters
OD	Outer diameter
OPD	Optical path difference
OTF	Optical transfer function
PCA	Principal component analysis
PCL	Plano-convex lens
PDB	PCA database
PML	Precision metrology lab
PSCCM	Point scanning chromatic confocal microscopy
PSF	Point spread function
PWM	Pulse width modulation
PZT	Piezoelectric translator
RCWA	Rigorous coupled wave analysis
RNN	Recurrent neural network
ROI	Region of interest
RSDI	Rayleigh-Sommerfeld diffraction integral
SDMH	Standard deviation of the measured height

SEM	Scanning electron microscopy
SIM	Structured illumination microscopy
SIMD	Single instruction multiple data
SNR	Signal-to-noise ratio
SPM	Scanning probe microscopy
SRC	Similarity response curve
SSE	Sum of squared error
STM	Scanning tunneling microscopy
TASC	Triangulation-assisted spectral confocal
TFM	Tilted flat mirror
WLI	White light interferometry
XP	Exit pupil
ZNCC	Zero mean NCC

Chapter 1
Optical Diffraction Theories

1.1 Introduction

This chapter introduces relevant background knowledge for modeling optical diffraction, which is integral to diffractive image microscopy. Simple examples are given for better comprehension.

1.2 Scalar Diffraction Theory

Light exhibits wave and particle behavior; in wave optics, it is treated as a wave. Wave optics accounts for wave effects such as diffraction and interference. Different types of radiation are propagated by different sources, such as sound energy being propagated by acoustic waves, radio energy being propagated by radio waves, and light energy being propagated by electromagnetic waves [1].

Diffraction refers to the optical behavior of light when some solid boundary, such as an aperture, confine it. When waves are disturbed by their boundaries, they do not follow a straight path. In our day-to-day experiences, we rarely notice the diffraction effect, as the effects of reflection and refraction are more commonly observed. However, diffraction becomes more significant when the structure size is comparable to the wavelength of light being used. Diffraction is crucial in any optical system where resolution is essential, such as astronomical telescopes, high-resolution microscopy, and photolithographic processes. In summary, understanding scalar diffraction is crucial to understanding the behavior of light in wave optics and its interaction with solid boundaries. Diffraction is often overlooked in everyday experiences due to the predominance of reflection and refraction effects, but its role becomes more critical when high-resolution measurements are required.

First, Maxwell's equations will be introduced, which set the foundations for characterizing the behaviors of electromagnetic fields. In MKS units, Maxwell's

L. Chen et al., *Diffractive Image Microscopy for 3D Imaging*, Springer Tracts in Mechanical Engineering, https://doi.org/10.1007/978-981-97-7782-2_1

equations in their differential form are given by Eqs. (1.1)–(1.4) [2]:

$$\nabla \times \vec{E} = -\mu \frac{\partial \vec{H}}{\partial t} \tag{1.1}$$

$$\nabla \times \vec{H} = \varepsilon \frac{\partial \vec{E}}{\partial t} \tag{1.2}$$

$$\nabla \cdot \varepsilon \vec{E} = 0 \tag{1.3}$$

$$\nabla \cdot \mu \vec{H} = 0 \tag{1.4}$$

In these equations, the electric field is represented by the vector quantity $\vec{E}$, which has three components (E_x, E_y, E_z), while the magnetic field is represented by the vector quantity $\vec{H}$, which has three components (H_x, H_y, H_z). The values of the permittivity and permeability of the medium through which the electromagnetic wave propagates are represented by ε and μ respectively. The representation of the cross-product of two vectors is given by "×" while the symbol represents the dot product of two vectors "·". These equations provide a framework for understanding the behavior of electromagnetic waves, including their propagation, reflection, and refraction.

In the case of light propagation through air, which is a nonmagnetic medium, the magnetic permeability is equal to the vacuum permeability, μ_0.

Applying mathematical operations in Maxwell equations, and if the medium is isotropic, linear, homogenous, and nondispersive, the electric field equation can be obtained as [3]:

$$\nabla^2 \vec{E} - \frac{n^2}{c^2} \frac{\partial^2 \vec{E}}{\partial t^2} = 0 \tag{1.5}$$

Here, n represents the refractive index of the medium, defined by Eq. (1.6).

$$n = \left(\frac{\varepsilon}{\varepsilon_0} \right)^{1/2} \tag{1.6}$$

The vacuum permittivity ε_0 and the velocity of propagation in vacuum c are given by Eq. (1.7).

$$c = \frac{1}{\sqrt{\mu_0 \varepsilon_0}} \tag{1.7}$$

Analogous to the electric field equation shown in Eq. (1.5), the magnetic field equation can be expressed as Eq. (1.8).

$$\nabla^2 \overrightarrow{H} - \frac{n^2}{c^2}\frac{\partial^2 \overrightarrow{H}}{\partial t^2} = 0 \tag{1.8}$$

The general equation of the scalar equation is expressed by Eq. (1.9), which $u(P, t)$ represents any scalar field component [4].

$$u(P, t) - \frac{n^2}{c^2}\frac{\partial^2 u(P, t)}{\partial t^2} = 0 \tag{1.9}$$

In summary, the equations above describe the fundamental behavior of electromagnetic waves during propagation through isotropic, linear, homogeneous, and nondispersive media such as air.

1.3 Monochromatic Field and Irradiance

In Eq. (1.10), a monochromatic scalar field propagating through an isotropic medium can be expressed as a cosine function with an amplitude $A(P)$ and a phase $\varphi(P)$ at position P. The optical frequency of the wave is denoted by υ. This equation denotes that the field has a periodic variation in time and space. The spatial distribution of the field is described by the function $u(P)$ in Eq. (1.10).

$$u(P, t) = A(P)\cos(2\pi\upsilon t - \varphi(P)) \tag{1.10}$$

Equation (1.11) provides a complex notation for the monochromatic scalar field. The fundamental part of this function, denoted by Re, gives us the actual physical value of the field at position P. This notation makes use of Euler's formula. The equation explains that the scalar field can be expressed as the product of a complex function $U(P)$ and an exponential function that depends on time. This complex notation is a useful tool for analyzing the spatial and temporal properties of the field.

$$u(P, t) = \mathrm{Re}\{U(P)\exp(-j2\pi\upsilon t)\} \tag{1.11}$$

In Eq. (1.12), the complex function $U(P)$ can be expressed in terms of its amplitude $A(P)$ and phase $\varphi(P)$. This equation tells us that the complex function combines a magnitude and a phase angle. The phase angle describes the relative timing of the wave at different positions in space.

$$U(P) = A(P)\exp[-j\varphi(P)] \tag{1.12}$$

By substituting Eq. (1.11) in Eqs. (1.9) and (1.13) can be obtained, which is a time-independent equation that must be satisfied by the function $U(P)$. This equation is a special case of the Helmholtz equation, which describes the propagation of waves in three-dimensional space. The wavenumber k is given by Eq. (1.14). This equation explains that the wavenumber is inversely proportional to the wavelength.

$$(\nabla^2 + k^2)U = 0 \tag{1.13}$$

where k is the wavenumber.

$$k = \frac{2\pi}{\lambda} \tag{1.14}$$

Finally, for a plane wave propagating in the Z direction, Eq. (1.15) denotes that the scalar field has a cosine variation in time and space with a wavenumber k and optical frequency υ. This equation is a simplified version of Eq. (1.10), where we assume that the wave travels in a single direction and is not affected by any obstacles.

$$u(Z, t) = A\cos(2\pi\upsilon t - kZ) \tag{1.15}$$

To further extend the previous Eq. (1.12), a more general equation is obtained as Eq. (1.16), where U_1 is the complex function that represents the spatial distribution of the field at a position (x, y) in the x–y plane, A_1 is the amplitude of the field, and $\varphi_1(x, y)$ is the phase of the field.

$$U_1(x, y) = A_1(x, y)\exp[-j\varphi_1(x, y)] \tag{1.16}$$

It is important to note that the field is located at some position '1' on the Z axis.

An optical detector responds to the time-averaged square magnitude of the field to detect the field. This means that the irradiance, denoted by I_1, is defined as the product of the field and its complex conjugate.

$$I_1 = U_1(x, y)U_1(x, y)^* \tag{1.17}$$

where "*" denotes the complex conjugate. Irradiance, in radiometry, refers to the flux (measured in watts per unit area), incident on the observation plane. It's a measure of power density and is commonly known as intensity. Thus, irradiance gives us an idea of how much power is present in a given area of the observation plane, and it is a critical parameter in many optical applications.

1.4 Rayleigh-Sommerfeld Diffraction Theory

Assuming the light propagated from a 2D source plane is determined by the coordinate variables ξ and η. The light is considered as monochromatic. The area Σ on the source plane defines the size of an illuminated pupil. Using first Rayleigh-Sommerfeld diffraction, the field in the observation plane can be predicted as $U_2(x, y)$, the equation for $U_2(x, y)$ is given by Eq. (1.18), where λ is the wavelength,

k is the wavenumber, Z represents the distance between two parallel coordinate systems, measured from the source centers to the center of the observation coordinate systems. r_{12} denotes the Euclidean distance between a position on the source plane and a position on the observation plane [2].

$$U_2(x, y) = \frac{Z}{j\lambda} \iint U_1(\xi, \eta) \frac{\exp(jkr_{12})}{r_{12}^2} \mathrm{d}\xi \mathrm{d}\eta \tag{1.18}$$

Since the source and observation locations are defined on parallel planes, the distance r_{12} can be calculated using Eq. (1.19). The distance r_{12} represents the distance between a point on the source plane and a point in the observation plane. The equation considers the distance Z between the source and observation planes and the distance between the coordinates x, y in the observation plane and the coordinates ξ, η in the source plane. Equation (1.19) predicts the field in the observation plane based on the field in the source plane and the distance between the two planes.

$$r_{12} = \sqrt{Z^2 + (x - \xi)^2 + (y - \eta)^2} \tag{1.19}$$

In the context of the propagation of light from a two-dimensional source plane, the expression given in Eq. (1.18) is a superposition integral, where the source and observation planes are on different planes (as shown in Fig. 1.1). This integral becomes a convolution integral, which is given in Eq. (1.20). Here, the Rayleigh-Sommerfeld impulse response, $h(x, y)$, is defined in Eq. (1.21), λ is the wavelength, ks the wavenumber for free space, and r is defined in Eq. (1.22).

$$U_2(x, y) = \iint U_1(\xi, \eta) h(x - \xi, y - \eta) \mathrm{d}\xi \mathrm{d}\eta \tag{1.20}$$

$$h(x, y) = \frac{Z}{j\lambda} \frac{\exp(jkr)}{r^2} \tag{1.21}$$

$$r = \sqrt{Z^2 + x^2 + y^2} \tag{1.22}$$

The Fourier transform of Eq. (1.20) is given in Eq. (1.23), where $\mathcal{F}$ is the Fourier transform operator and $\mathcal{F}^{-1}$ is its inverse. H is the transfer function of Rayleigh-Sommerfeld given in Eq. (1.24). f_X and f_Y are spatial frequencies and λ is the cutoff wavelength given in Eq. (1.25) [2].

$$U_2(x, y) = \mathcal{F}^{-1}\{\mathcal{F}\{U_1(x, y)\}\mathcal{F}\{h(x, y)\}\} \tag{1.23}$$

$$U_2(x, y) = \mathcal{F}^{-1}\{\mathcal{F}\{U_1(x, y)\}H(f_X, f_Y)\} \tag{1.24}$$

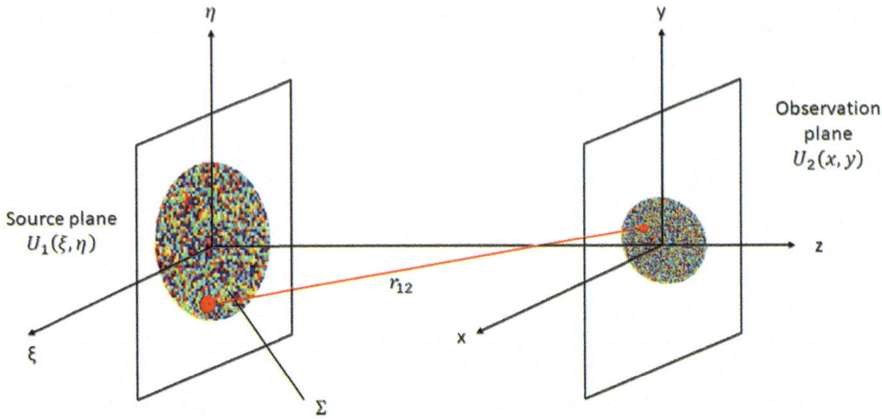

Fig. 1.1 Light propagation from source plane to observation plane

$$H(f_X, f_Y) = \begin{cases} \exp\left(jkz\sqrt{1-(\lambda f_X)^2-(\lambda f_Y)^2}\right), & \sqrt{f_X^2+f_Y^2} < \frac{1}{\lambda} \\ 0 & \text{otherwise} \end{cases} \tag{1.25}$$

1.5 Fresnel Approximation

The Fresnel diffraction equation, given in Eq. (1.26), is a simplified version of the Kirchhoff-Fresnel diffraction that applies to the propagation of waves in the near field. This equation assumes that the rays of light that pass through the aperture or around the object make only a tiny angle concerning the optical axis, known as the paraxial approximation.

To reduce the complexity of the term given in Eq. (1.18), binomial expansion is used to approximate the term, neglecting the minor term. The Fresnel diffraction equation can be simplified by approximating the distance term, as shown in Eq. (1.26). This equation is a convolution integral, similar to Eq. (1.18). The impulse response function for this equation is given by Eq. (1.27), and the transfer function is given by Eq. (1.28) [2].

$$U_2(x, y) = \frac{e^{jkz}}{j\lambda z} \iint U_1(\xi, \eta) \exp\left\{j\frac{k}{2z}\left[(x-\xi)^2 + (y-\eta)^2\right]\right\} d\xi d\eta \tag{1.26}$$

$$h(x, y) = \frac{ze^{jkz}}{j\lambda z} \exp\left[\frac{jk}{2z}(x^2+y^2)\right] \tag{1.27}$$

$$H(f_X, f_Y) = e^{jkz} \exp\left[j\pi\lambda z\left(f_X^2+f_Y^2\right)\right] \tag{1.28}$$

Notably, the Fresnel approximation applies only to wave propagation in the near field, where the distance between the source and observation planes is relatively small. The Kirchhoff-Fresnel diffraction is a more general formula applicable in both the near and far fields, but it is more complex to calculate.

1.6 Fraunhofer Approximation

When light propagates far away from the aperture in the source plane, the optical field in the observation plane can be described using the Fraunhofer diffraction integral. The Fraunhofer diffraction equation, which approximates the field in the observation plane, is given by Eq. (1.29). This equation represents the diffraction pattern that results from the interference of light waves passing through an aperture and propagating to a distant observation plane [2].

$$U_2(x, y) = \frac{\exp(jkZ)}{j\lambda Z} \exp\left[j\frac{k}{2Z}(x^2 + y^2)\right] \iint U_1(\xi, \eta) \exp\left[-j\frac{2\pi}{\lambda Z}(x\xi + y\eta)\right] \mathrm{d}\xi \mathrm{d}\eta \tag{1.29}$$

The Fraunhofer diffraction equation is essentially a Fourier transform of the source field with the substitutions Eq. (1.30). This equation can determine the intensity and phase of the diffraction pattern produced by a given aperture. It can help to determine the spot size and cross-section of the pattern.

$$f_\xi \to \frac{x}{\lambda z}, f_\eta \to \frac{y}{\lambda z} \tag{1.30}$$

For instance, consider an aperture of size 1 mm and a wavelength of light of 550 nm. Suppose the observation plane is located at a distance of 70 m from the aperture. Using the Fraunhofer diffraction equation, we can calculate the pattern's spot size and cross-section, as shown in Fig. 1.2.

1.7 Aberrations: Seidal Polynomials

Any real optical system has optical aberration, which means the actual wavefront that propagates in the optical system deviates from their ideal forms, say perfect plane waves or spherical waves, which can be seen in Fig. 1.3.

Seidel aberrations are a set of five primary monochromatic aberrations that occur in optical systems. These include coma, spherical aberration, astigmatism, distortion, and Petzval field curvature. They are named after Ludwig von Seidel, who was the first to derive them mathematically [5].

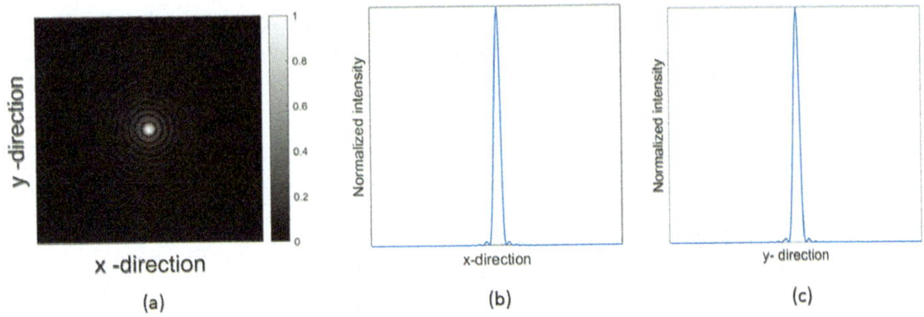

Fig. 1.2 **a** Fraunhofer intensity distribution on image plane; **b** x-axis cross-section of intensity distribution; **c** y-axis cross-section of intensity distribution

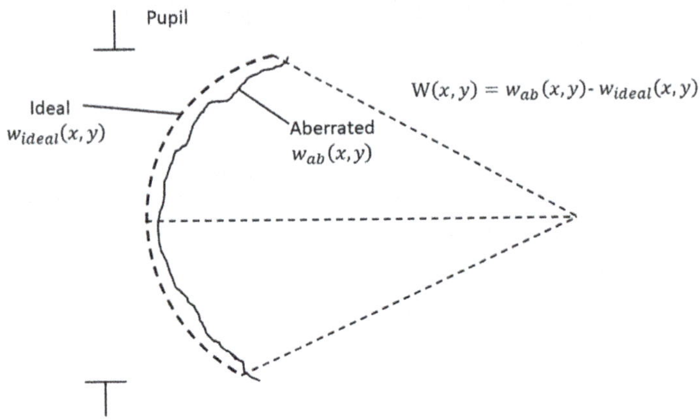

Fig. 1.3 Aberration effect on wavefront

Spherical aberration arises when light rays passing through the edge of a lens or mirror converge at a distinct point compared to those passing through the center.

This aberration is commonly denoted by W_{040}. Coma is the result of focusing light rays from an off-axis point source. This type of aberration is commonly denoted by W_{131}. Astigmatism occurs when light rays from a single point are focused into two lines at right angles. This type of aberration is commonly denoted by W_{222}. Petzval field curvature occurs when the image formed by a lens or mirror is curved rather than flat. This type of aberration is commonly denoted by W_{220}. Distortion occurs when the magnification of an image changes across the field of view. This type of aberration is commonly denoted by W_{311}. Defocus is another type of optical aberration that occurs when the image plane is not located at the proper distance from the lens or mirror. It can be thought of as a shift in the position of the image plane along the optical axis. This type of aberration is commonly denoted by W_d. A common form applied in imaging systems is described by [6, 7].

$$W(\hat{u}_0; \rho; \theta) = \sum_{j,m,n} W_{k\,l\,m} \hat{u}_0^k \rho^l \cos^m \theta; \; k = 2j + m, \; l = 2n + m \tag{1.31}$$

where ρ and θ represent the normalized radial distance and the angle within the exit plane, respectively. The above equation is in polar coordinates. Hence, converting it to a Cartesian coordinate for calculation is convenient. Here is the converted Cartesian coordinates form:

$$\rho = \sqrt{\hat{x}^2 + \hat{y}^2} \;\; \text{and} \; \rho \cos \theta = \hat{x} \tag{1.32}$$

Hence (Eq. 3.34) can then be written in a Cartesian form as [6, 7]:

$$\begin{aligned} W(\hat{u}_0; \hat{x}, \hat{y}) = {} & W_d(\hat{x}^2 + \hat{y}^2) + W_{040}(\hat{x}^2 + \hat{y}^2)^2 + W_{131}\hat{u}_0(\hat{x}^2 + \hat{y}^2)\hat{x} + W_{222}\hat{u}_0^2\hat{x}^2 \\ & + W_{220}\hat{u}_0^2(\hat{x}^2 + \hat{y}^2) + W_{311}\hat{u}_0^3\hat{x} \end{aligned} \tag{1.33}$$

All the kinds of Seidel aberration have been plotted in Fig. 1.4.

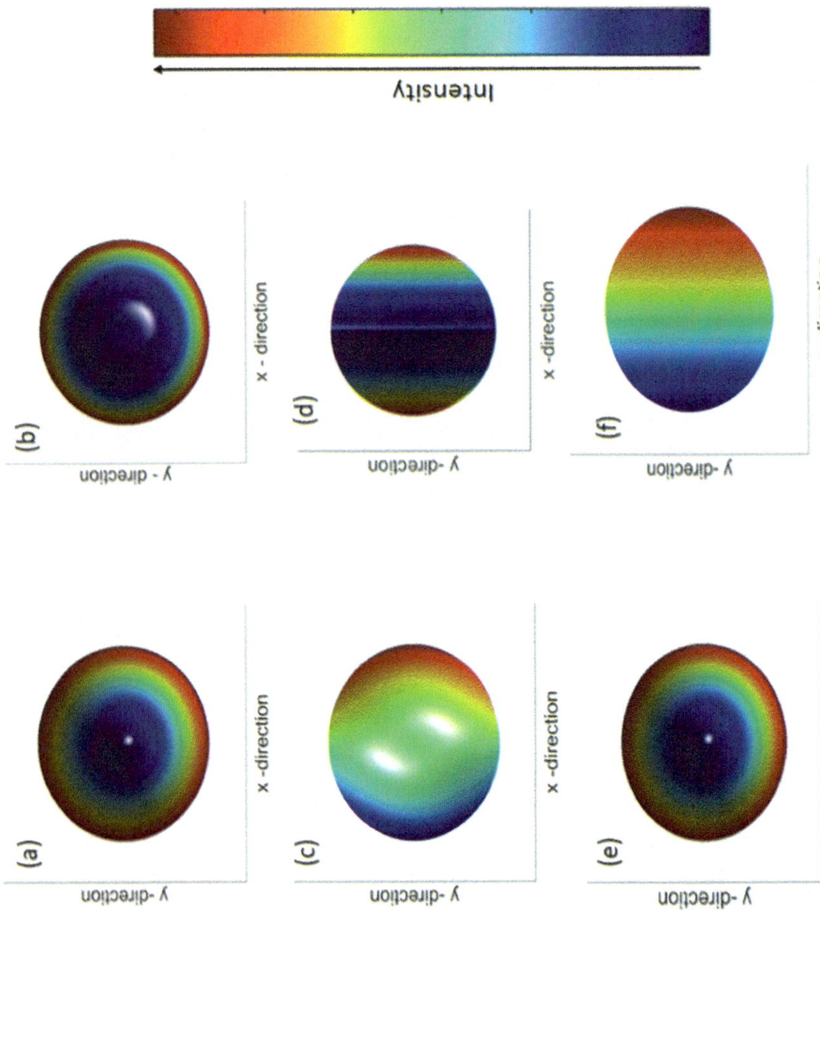

Fig. 1.4 Seidel aberration plot **a** defocus, **b** spherical, **c** coma, **d** astigmatism, **e** field curvature, **f** distortion

References

1. Born M, Wolf E (2013) Principles of optics: electromagnetic theory of propagation, interference and diffraction of light. Elsevier
2. De Araújo CB, Gomes AS, Boudebs G (2016) Techniques for nonlinear optical characterization of materials: a review. Rep Prog Phys 79(3):036401
3. Hecht E (2012) Optics. Pearson Education India
4. Mahajan VN (1998) Optical imaging and aberrations. Optical imaging and aberrations/Virendra N. Mahajan. Bellingham
5. Fischer RE et al (2000) Optical system design, vol 599. Citeseer
6. Wyant JC, Creath K (1992) Basic wavefront aberration theory for optical metrology. Appl Opt Opt Eng 11(part 2):28–39
7. Geary JM (2002) Introduction to lens design with practical ZEMAX® examples. Willmann-Bell. Inc., Richmond

Chapter 2
Optical Imaging Theories

2.1 Introduction

Although a real optical system is usually more complex, critical properties of interest can often be studied via a representative physical model. This chapter proposes a physical model to describe the image formation in an optical microscope. The optical system is expressed as a "pupil function" in the proposed model. The pupil function captures changes in amplitude and phase as light passes through it. Examples of point spread function simulation with the proposed model are demonstrated.

2.2 Geometric Optics

Geometrical optics is a fundamental branch of optics that provides a simplified framework for understanding the behavior of light. Instead of considering the wave nature of light, geometrical optics focuses on the propagation of light using rays. This simplified model assumes that light travels in straight lines and experiences reflection and refraction at interfaces between different materials. We can analyze and design various optical systems using geometrical optics, including lenses, mirrors, and prisms. It serves as a valuable tool for studying the principles of light propagation and optimizing the performance of optical devices.

Geometrical optics provides a simplified yet useful framework for studying the behavior of light. In this approach, light is represented as a collection of rays that travel in straight lines and interact with various optical elements such as lenses, mirrors, and prisms. When a ray encounters an optical boundary where the refractive indices of the media differ, it can undergo reflection, refraction, or both.

The focal length in geometrical optics represents the distance between a mirror or lens and the point where collimated rays converge or appear to diverge. It is a

L. Chen et al., *Diffractive Image Microscopy for 3D Imaging*, Springer Tracts in Mechanical Engineering, https://doi.org/10.1007/978-981-97-7782-2_2

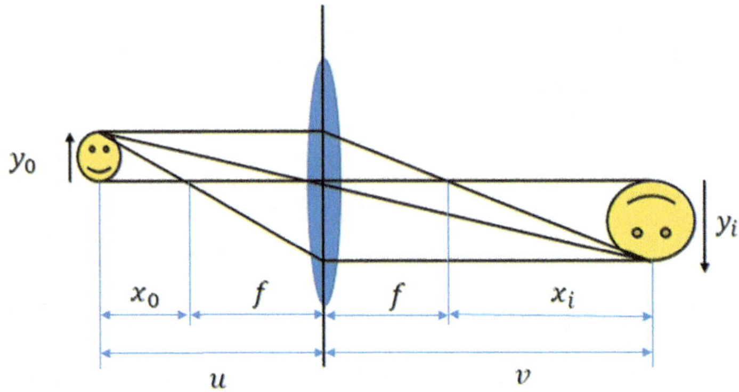

Fig. 2.1 Image formation by a thin lens

critical parameter for designing optical systems, and the curvature of the lens or mirror surface primarily determines it.

In optical systems, achieving the best focus for imaging is often desirable. Under paraxial conditions, where the ray angles are small or close to the optical axis, the relationship required for optimal focus imaging can be described by the lens equation [1].

For the specific image situation depicted in Fig. 2.1, image formation by a thin lens, the lens equation (Eq. 2.1) in its Gaussian form provides the relationship needed for achieving the best focus under the given paraxial conditions. This encapsulates the principles necessary for precise imaging in optical systems.

$$\frac{1}{u} + \frac{1}{v} = \frac{1}{f} \tag{2.1}$$

where u denotes the object distance, v represents the image distance, and f indicates the focal length.

In geometrical optics, lateral or transverse magnification holds significant importance. It characterizes the relationship between the height of the image formed and the height of an object. This magnification is calculated by taking the scale of the ratio between the image height (y_i) to the object height (y_o) given in Eq. (2.2) [1].

$$M_t = \frac{y_i}{y_o} = -\frac{v}{u} \tag{2.2}$$

The resulting value of the magnification can be negative, indicating an inverted image formation. In such cases, the image appears upside down compared to the object. Conversely, a positive magnification signifies an upright image.

The pupil of an optical system serves as the aperture through which light traverses, playing a crucial role in the overall system behavior. It can be defined as the image of the entrance pupil created by the optical system. The entrance pupil represents the physical aperture through which light enters the system, while the exit pupil corresponds to the virtual aperture through which light appears to exit the system.

In the design of optical systems, the entrance pupil (E_{EP}) holds significant importance as it governs the amount of light that can enter the system and the angles at which the light rays impinge upon the optical elements. Various factors, such as the physical aperture of the lens or mirror, can determine the size of the entrance pupil. Alternatively, it may be dictated by the dimensions of a diaphragm or iris in front of the lens.

On the other hand, the exit pupil (E_{XP}) is influenced by the system's optical characteristics, including the image's magnification and position. Serving as a virtual aperture, the exit pupil determines the angles at which light rays leave the system.

To regulate the amount of light entering the system, aperture stops (A.S) are employed. These stops can be physical or virtual diaphragms that limit the size of the entrance pupil. Adjusting the aperture stop, such as a physical diaphragm in a camera lens or a circular opening at the telescope's entrance, can effectively control the amount of light entering the system.

Understanding the concepts of entrance and exit pupils and the role of aperture stops is crucial for designing and optimizing optical systems across various applications.

In optical systems, the concepts of the chief and marginal rays play a vital role in assessing system performance.

The chief ray refers to a specific ray of light that traverses through the center of the entrance pupil within an optical system. It serves as a fundamental reference ray for understanding the behavior of other rays within the system. Its significance lies in the fact that it passes through the center of the system without being deflected by any optical elements. Consequently, the chief ray provides a precise reference point for determining the size and location of the image formed by the system.

Conversely, the marginal ray represents a ray of light that passes through the outer edge of the entrance pupil. Similar to the chief ray, it serves as a reference for analyzing the behavior of other rays within the system. However, unlike the chief ray, the marginal ray is subject to deflection by the optical elements present in the system. By observing the path of the marginal ray, valuable insights can be gained regarding the system's distortion and aberrations, enabling an assessment of the image quality produced by the system [1].

Together, the chief and marginal rays provide essential information about the behavior of light as it traverses an optical system. Tracing these rays throughout the system makes it possible to determine the image location, size, distortion, and aberrations. Consequently, the chief and marginal rays serve as invaluable tools for analyzing the performance of optical systems and optimizing their design [2].

To illustrate the propagation of rays within a two-lens optical system, a comprehensive diagram has been included in Fig. 2.2. This diagram showcases the various types of rays and stops, aiding in visualizing the system and its ray propagation characteristics.

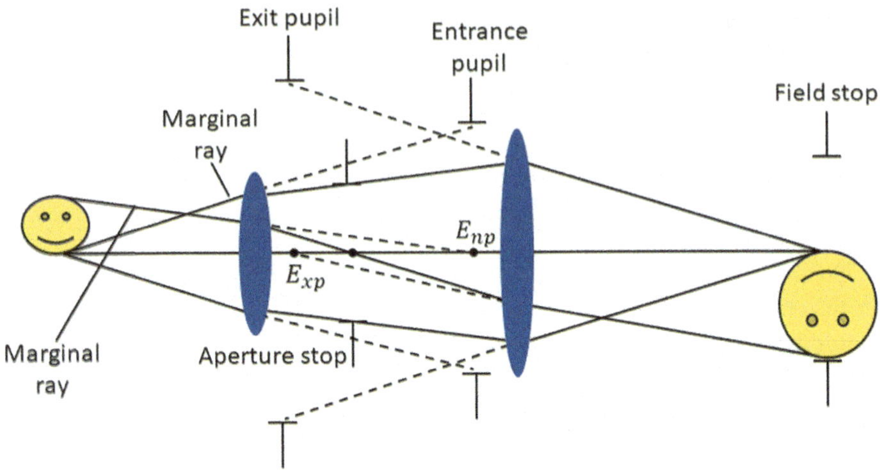

Fig. 2.2 A generalized two-lens optical system

The f-number is a critical parameter in optics that plays a crucial role in determining the performance of an optical system, especially concerning its light-capturing capacity and depth of field. This is defined as the ratio of the focal length of a lens to its effective aperture diameter, and the f-number holds great importance as it determines the amount of light that can pass through the system. A smaller f-number indicates a larger effective aperture diameter, allowing more light to traverse the optical system. This feature proves valuable in low-light scenarios, where a larger aperture can gather ample light and produce a brighter image.

Furthermore, the f-number directly influences the depth of field of the system. A smaller f-number (larger aperture) results in a shallower depth of field, causing only a small portion of the image to remain in focus while the remaining areas appear blurred. This effect can be advantageous in portrait or macro photography, where a shallow depth of field effectively isolates the subject from the background, yielding aesthetically pleasing images [3].

While geometrical optics is a useful tool for assessing and designing optical systems, it has certain limitations. One major constraint is its inability to account for the wave nature of light is responsible for various phenomena, including diffraction and interference. Additionally, geometrical optics is only valid for describing light behavior in regions where the system's scale significantly exceeds the wavelength of light.

Notwithstanding these limitations, geometrical optics remains an indispensable and widely utilized tool in optics. It offers a straightforward and intuitive approach to analyzing and designing optical systems, serving as the foundation for more advanced models considering light's wave nature.

2.3 Diffraction Optics

Figure 2.3 shows a visual representation demonstrating that the physical components of an optical system, such as lenses, mirrors, and irises, can be simplified into models based on the entrance and exit pupils (EP and XP). In this context, the distance from an object point along the optical axis toward the entrance pupil is denoted as Z_{EP}, while the distance from the exit pupil to an image point on the optical axis is denoted as Z_{XP}. D_{EP} and D_{XP} are the diameters of the entrance and exit pupils, respectively in Fig. 2.3.

In Fourier optics, the general arrangement for imaging considerations is depicted in the accompanying Fig. 2.4. Within this setup, the object field is symbolized as $U_O(x, y)$, indicating the distribution of the object's spatial properties.

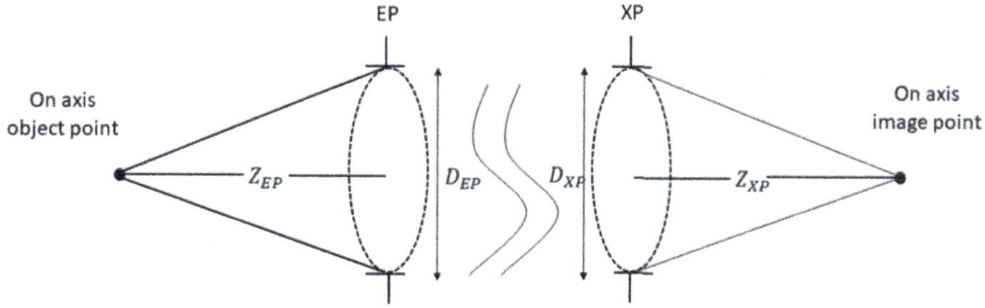

Fig. 2.3 Generalized optical model

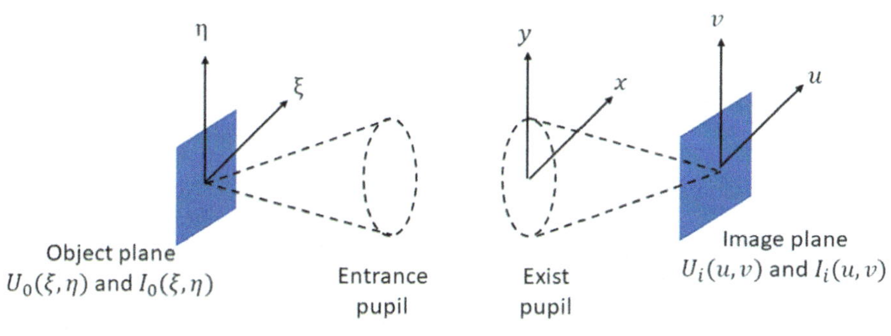

Fig. 2.4 Imaging coordinate definition

2.4 Coherent Imaging Theory

Coherent and incoherent imaging are the fundamental concepts in optics that explain the behavior of light as it traverses an optical system and generates an image. Coherent imaging involves using light from a single source, such as a laser. In this scenario, all the light waves have a consistent phase relationship, resulting in predictable interference patterns. This coherence enables the formation of detailed and high-resolution images as the light waves interact with the object in a well-defined manner. The coherent imaging process is mathematically described as the convolution of the optical field, as shown in Eq. (2.3). The variables u and v represent the coordinates within the image plane. U_i represents the light field in the image plane, and the coherent response of the optical system is denoted by h. U_g corresponds to the geometrical optics-based ideal image field, which corresponds to a magnification of the object field $U_0(x, y)$.

The magnification factor M_t is an absolute value, Eq. (2.4) [4].

$$U_\mathrm{i}(u, v) = h(u, v) \otimes U_\mathrm{g}(u, v) \tag{2.3}$$

$$U_\mathrm{g}(u, v) = \frac{1}{|M_\mathrm{t}|} U_0\left(\frac{u}{M_\mathrm{t}}, \frac{v}{M_\mathrm{t}}\right) \tag{2.4}$$

The equation describes the spatial coordinates, but a Fourier transform is applied to the equation to analyze the system in the frequency domain. The resulting spectra for the equation in the frequency domain are represented in the corresponding Eq. (2.5).

$$G_\mathrm{i}(f_U, f_V) = H(f_U, f_V) G_\mathrm{g}(f_U, f_V) \tag{2.5}$$

$$H(f_U, f_V) = P(-\lambda Z_\mathrm{XP} f_U, -\lambda Z_\mathrm{XP} f_V) \tag{2.6}$$

where P is the pupil function of the system.

Lenses are commonly used in optical components and typically have a circular shape. For illustrative purposes, a diffraction-limited circular pupil function is considered, as defined in Eq. (2.7). Here, $R_\mathrm{XP} = D_\mathrm{XP}/2$ is equal to half the exit pupil diameter D_{XP}. Based on this, the coherent transfer function is defined as shown in Eq. (2.8), which can also be expressed in the subsequent Eq. (2.9) [4].

$$P(x, y) = \mathrm{circ}\left(\frac{\sqrt{(x^2 + y^2)}}{R_\mathrm{XP}}\right) \tag{2.7}$$

$$H(f_U, f_V) = \mathrm{circ}\left(\frac{\sqrt{(-\lambda Z_\mathrm{XP} f_U)^2 + (-\lambda Z_\mathrm{XP} f_V)^2}}{R_\mathrm{XP}}\right) \tag{2.8}$$

$$H(f_U, f_V) = \mathrm{circ}\left(\frac{\sqrt{(f_U)^2 + (f_V)^2}}{f_0}\right) \tag{2.9}$$

The coherent transfer function incorporates the coherent cutoff frequency, f_0, for the circular aperture, which is given as follows:

$$f_0 = \frac{R_{XP}}{\lambda Z_{XP}} \tag{2.10}$$

where R_{XP} is the exit pupil radius, λ is the wavelength of the light source, and Z_{XP} is the exit pupil distance.

2.5 Incoherent Imaging Theory

Incoherent imaging refers to imaging an object using light emitted from a non-coherent source, like the sun or a light bulb. In this case, the light waves are not in phase with each other, meaning they have a random and unpredictable interference pattern. This lack of coherence can result in lower-resolution images, as the light waves interact with the object in a more chaotic way [5].

Considering the incoherent light source that illuminates the object, the linear space invariant model will be given in Eq. (2.11) for imaging [6].

$$I_i(u, v) = |h(u, v)|^2 \otimes I_g(u, v) \tag{2.11}$$

As described earlier, the variable h represents the coherent impulse response while I_g presents the ideal geometric irradiance image. It is worth mentioning that coherent imaging exhibits linearity in relation to the field, while incoherent imaging demonstrates linearity with the irradiance. The term $|h(u, v)|^2$ relates to the impulse response, commonly called the PSF.

The corresponding spectra of the function given in the equation can be transformed to the frequency domain by Eq. (2.12).

$$G_i(f_U, f_V) = \mathcal{H}(f_U, f_V) G_g(f_U, f_V) \tag{2.12}$$

where $\mathcal{H}$ is known as the OTF. The OTF represents a normalized autocorrelation of the coherent transfer function $H(f_U, f_V)$.

2.6 Pupil Function

The pupil function is a fundamental concept in optics that describes the transmission properties of an optical system. It is a mathematical function that describes how a light wave front is transmitted through an aperture or lens, and it is used to calculate the image formation in the system [5].

The aperture function is a binary function that describes which parts of the incoming light are transmitted and which parts are blocked by the optical system. The Fourier transform of the aperture function gives the complex transmission function, which describes the amplitude and phase of the transmitted light as a function of position in the pupil plane.

The pupil function is important in a wide range of optical applications, including microscopy, lithography, and astronomy. In microscopy, the pupil function is used to describe the diffraction-limited resolution of the microscope and to optimize the design of the microscope objective and other optical components. In lithography, the pupil function is used to optimize the exposure and resolution of the printed pattern on a substrate. In astronomy, the pupil function is used to describe the transmission properties of the telescope and optimize the image quality.

By employing the complex phasor method, the distorted pupil function can be defined as follows [4]:

$$P(x, y) = \mathrm{circ}\left(\frac{\sqrt{x^2 + y^2}}{R_{\mathrm{XP}}}\right) \exp\left(-jkW\left(\frac{x}{R_{\mathrm{XP}}}, \frac{y}{R_{\mathrm{XP}}}\right)\right) \tag{2.13}$$

where circ describes the diffraction-limited exist pupil, R_{XP} is the exit pupil radius.

2.7 Imaging Transfer Function (ITF) and Coherent Transfer Function (CTF)

ITF is a mathematical tool used in optics to describe the transfer of information from an object to an image. It relates the spatial frequencies of the object to the spatial frequencies of the image, and it is often used to characterize the resolution and contrast of an imaging system. The ITF definition involves the Fourier transform of the PSF of the optical system. The PSF describes the system response of a point source or an infinitely small object, and it determines the blurring or spreading of the image due to diffraction, aberrations, and other factors. The relationship between the ITF and the pupil function of the imaging system is fundamental to the analysis of image formation. The pupil function describes the transmission properties of the optical system, and it determines the diffraction pattern formed by the system. The diffraction pattern, in turn, affects the system's PSF and ITF [5].

The CTF is a mathematical tool used in microscopy to describe the transfer of information from the specimen to the image, considering the illumination's coherence and incoherence. It is directly related to the optical system's pupil function and ITF. The ITF is the Fourier transform of the PSF, characterizing the transfer of information from the specimen to the image. It is related to the CTF through convolution with the aperture function of the imaging system and considers the coherence and incoherence of the illumination.

In practice, the CTF can be determined from the pupil function by performing the Fourier transform of the complex conjugate of the pupil function. This calculation considers the transfer of spatial frequency components of the object through the imaging system and allows for the calculation of the resolution and contrast of the final image.

The relationship between the pupil function, the PSF, the CTF, and the ITF is crucial for understanding the imaging properties of an optical system. By manipulating the pupil function, it is possible to control the resolution, contrast, and depth of field of the system, and to optimize it for specific applications. The ITF can be used to quantify an image's information content and optimize the imaging conditions for maximum information transfer. The equation given was for the optical system without any aberration. The coherent transfer function for the optical system can be written in the same way by including normalized image coordinates and is given as [6]:

$$H(f_U, f_V) = P(-\lambda Z_{\mathrm{XP}} f_U, -\lambda Z_{\mathrm{XP}} f_V) \tag{2.14}$$

2.8 Point Spread Function

The point spread function (PSF) characterizes the system's response to an ideal point source. In other words, it represents the image produced by a point source of light after it has passed through the optical system. The PSF is a fundamental concept in imaging, as it describes an optical system's fundamental limits of resolution.

The relationship between the PSF and the CTF can be derived from the Fourier properties of the convolution operation. The PSF is the inverse Fourier transform of the CTF. The CTF, on the other hand, is the Fourier transform of the autocorrelation of the PSF. Using the Fourier transform properties, it can be shown that the PSF equals the inverse Fourier transform of the CTF, Eq. (2.15) [7].

$$|h(u, v)|^2 = \left|\mathcal{F}^{-1}\{H(f_U, f_V)\}\right|^2 \tag{2.15}$$

In practical terms, the PSF evaluates the imaging system's resolution and contrast. The CTF, on the other hand, measures the system's frequency response and describes how well the system can resolve different spatial frequencies.

2.9 Model Validation

The equations explained above are interrelated for modeling the optical system. Initially, an example is used to demonstrate the optical model using these equations. Subsequently, these concepts are applied to the proposed optical system to derive the optical wave model.

A plano-convex lens (110-0243E) from EKSMA OPTICS has been chosen for demonstration. The company provides the design prescription data for the lens at a wavelength of 546 nm as a Zemax file. This file indicates that the entrance and exit pupil diameters are 22.86 mm, with an effective focal length of 400.0187 mm. The paraxial working f/# is 17.49863, and the exit pupil position from the image plane is -399.7817 mm. The negative sign indicates the position of the exit pupil relative to the image plane. The back focal length is 398.0433 mm with a center thickness of 3 mm. The front radius of curvature of the lens is 207.5 mm, while the back surface is flat due to the plano-convex design. The lens material is N-BK7.

Initially, no aberrations are considered in the model, and the results are compared with those from Zemax. Later, aberrations are optimized, and the comparison with the Zemax results is repeated. It is important to note that the modeling is done only on-axis. The ray-tracing diagram of the lens is shown in Fig. 2.5. The PSF results from the model and its cross-sectional view in the u and v directions are shown in Fig. 2.6.

The plot of PSF and its cross-sectional view from Zemax has been shown in Fig. 2.7.

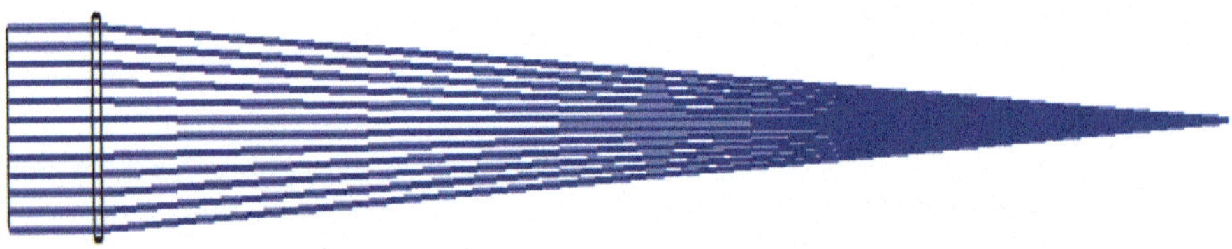

Fig. 2.5 Ray tracing diagram of 110-0243E

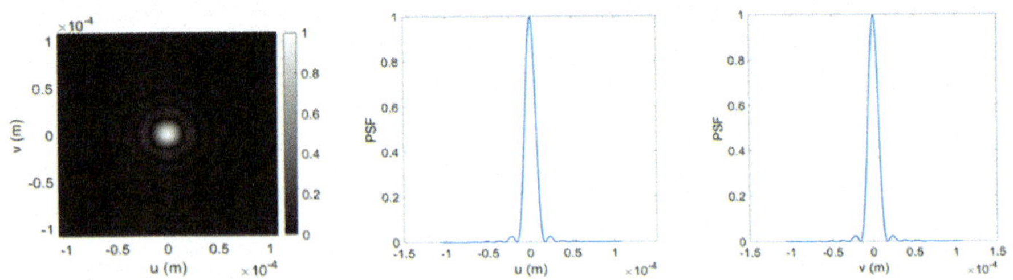

Fig. 2.6 PSF and its cross-sectional plot in the u and v direction obtained from the mathematical model

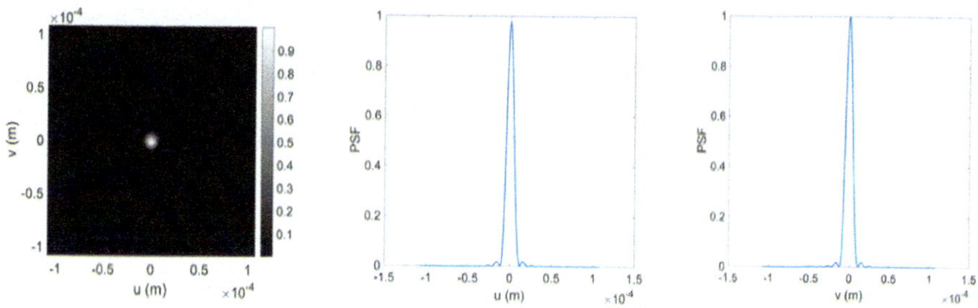

Fig. 2.7 PSF and its cross-sectional plot in the *u* and *v* direction obtained from Zemax

Since the PSF plotted in the figure does not consider any aberration, an optimization approach can be used to determine the aberration coefficient value. Common optimization approaches include fminsearch, fmincon, and Bayesian Adaptive Direct Search (BADS) [8].

Optimization processes like fminsearch, fmincon, and BADS are used to minimize aberrations, which measure the deviation of an optical system from ideal behavior. fminsearch is a built-in function in MATLAB that uses the Nelder-Mead simplex algorithm to minimize a given function. It works by starting with a simplex (a set of points) in the parameter space and iteratively moving and reflecting the points to converge toward the function's minimum. fminsearch is suitable for unconstrained optimization problems. However, it may not perform well for problems with many parameters or when the function to be optimized is noisy or has many local minima.

fmincon is another built-in function in MATLAB that can handle constrained optimization problems. It uses the interior-point algorithm to find the minimum of a given function subject to constraints. The constraints can be linear or nonlinear and can be defined as equalities or inequalities. fmincon works well for problems with constraints but can be slow for problems with many parameters.

BADS is an optimization algorithm that combines Bayesian optimization and direct search. It works by iteratively building a probabilistic model of the objective function and using it to guide the search toward promising regions of the parameter space. BADS can handle problems with many parameters, noisy functions, and constraints. It is also efficient in terms of the number of function evaluations required to find a good solution.

In the context of aberration calculation, BADS is superior to fminsearch and fmincon because it can handle problems with many parameters and constraints, which are common in optical design. It is also robust to noisy functions, which is important when dealing with experimental data. Moreover, BADS is efficient in terms of the number of function evaluations required to find a good solution, saving time and computational resources.

In summary, fminsearch, fmincon, and BADS are optimization algorithms used to minimize the aberration of an optical system. While fminsearch and fmincon are built-in MATLAB functions suitable for unconstrained and constrained optimization problems, BADS is a more advanced optimization algorithm that can handle problems

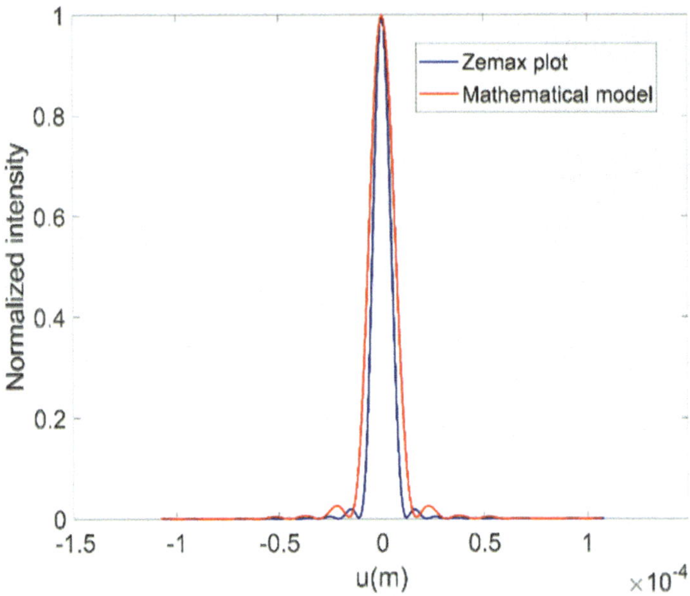

Fig. 2.8 Comparison plot of mathematical model and Zemax

with many parameters, constraints, and noisy functions. Therefore, BADS is superior to fminsearch and fmincon for aberration calculation in optical design.

In the analysis, we focus solely on the optical axis, which implies the presence of only spherical aberration. The BADS optimization method is used to compute the aberration term and compare it to the Zemax result. Figure 2.8 displays a comparison between the Zemax plot and the fitted curve derived from our model, which incorporates the spherical aberration. The sum of squared errors is used as the loss function during the optimization process. Our optimization process yields a spherical aberration value of 0.1927, relatively close to the 0.1328 value generated by Zemax. Consequently, the BADS optimization method can be adopted to calculate the final aberration term in our proposed model.

In summary, a mathematical model and optimization process have been discussed and used for the proposed model. An example was taken for validation, and the result was compared with the Zemax result.

References

1. Hecht E (2012) Optics. Pearson Education India
2. Mahajan VN (2013) Optical imaging and aberrations, Part III: wavefront analysis
3. Fischer RE et al (2000) Optical system design, vol 599. Citeseer
4. Goodman JW (2005) Introduction to Fourier optics. Roberts and Company Publishers

5. Born M, Wolf E (2013) Principles of optics: electromagnetic theory of propagation, interference and diffraction of light. Elsevier
6. Mahajan VN (1998) Optical imaging and aberrations. Optical imaging and aberrations/Virendra N. Mahajan. Bellingham
7. Smith WJ (2008) Modern optical engineering: the design of optical systems. McGraw-Hill Education
8. Acerbi L, Ma WJ (2017) Practical Bayesian optimization for model fitting with Bayesian adaptive direct search. Adv Neural Inf Process Syst 30

Chapter 3
State-of-the-Art Optical Profilometry Techniques

3.1 Introduction

This chapter reviews state-of-the-art optical profilometry techniques and discusses surface-dependent errors that arise during profile measurements. Potential methods developed for minimizing these errors are examined to highlight the scientific problem to be solved by diffractive image microscopy.

3.2 Monochromatic Confocal Microscopy

Monochromatic confocal microscopy (MCM) is a fundamental optical microscopy technique. Its core principle combines point illumination and point detection, achieved using two spatial filters, typically circular pinholes, as shown in Fig. 3.1. The illumination pinhole limits the irradiation region to a tiny spot on the detected object, minimizing the influence of energy from the surrounding area on the final signal. Consequently, the final signal is dominated by the physical information within the tiny spot. After illumination, the light reflected or scattered by the object is collected by the optical system and focused near the detection pinhole. This further restricts the observed region by blocking off light outside the detected point. Light can only pass through the detection pinhole efficiently when the detected surface is in focus. An intensity sensor placed after the detection pinhole detects the passing light energy. With vertical scanning, the position with the strongest signal indicates the position of the detected point. This results in a point-wise detection feature with a sharp depth response curve (DRC) and a unique three-dimensional optical sectioning property [1–4].

In general, when the optical system is not aberration-free, the measured profile can be affected by the local surface slope. As illustrated in Fig. 3.2, the surface slope can influence the region where signal light passes through the system pupil. Therefore,

L. Chen et al., *Diffractive Image Microscopy for 3D Imaging*, Springer Tracts in Mechanical Engineering, https://doi.org/10.1007/978-981-97-7782-2_3

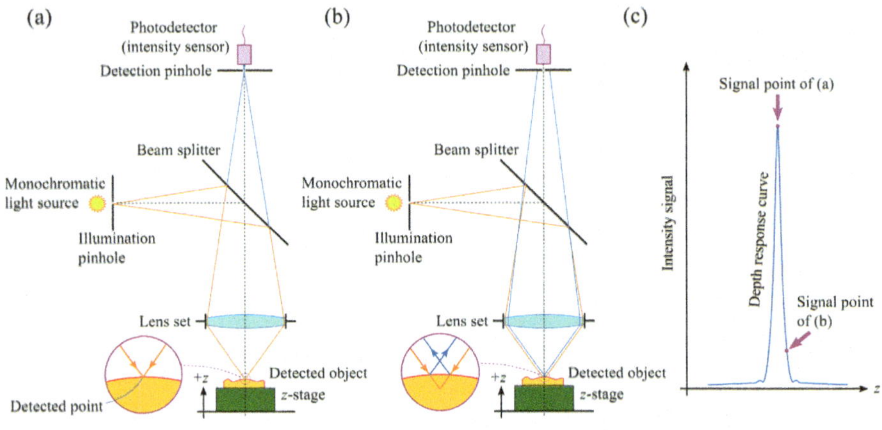

Fig. 3.1 Principle of monochromatic confocal microscopy

the light field that reaches the detection pinhole can vary with the surface slope, even if the longitudinal surface position is maintained. Consequently, the final signal is distorted, leading to systematic error. Extensive research in this area indicates that if the profile of the detected object is known in advance (e.g., a standard reference ball), proper compensation can significantly reduce this error. However, this is not generally applicable since prior knowledge cannot be provided for measuring an unknown object.

As shown in Fig. 3.3, another study reports that a curved surface can lead to artifacts in the measured profile [6]. This occurs because a locally curved surface reflects the incoming wavefront differently than a flat surface, resulting in a change in the shape of the reflected wavefront. Consequently, the final signal is distorted, leading to measurement error. The authors proposed a more advanced model to

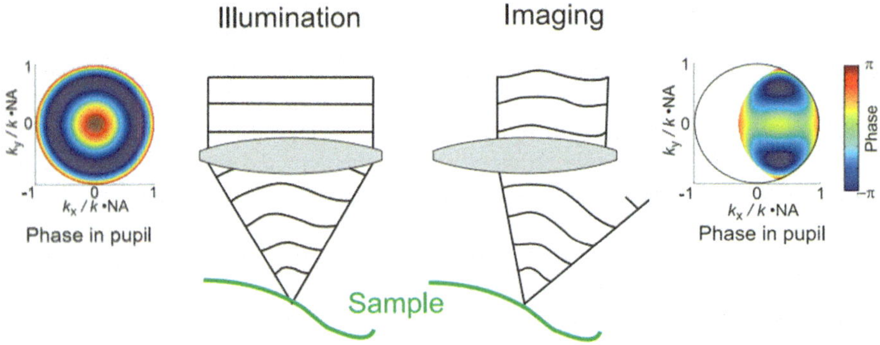

Fig. 3.2 A tilted local surface can affect the region where signal light passes through the system pupil [5]

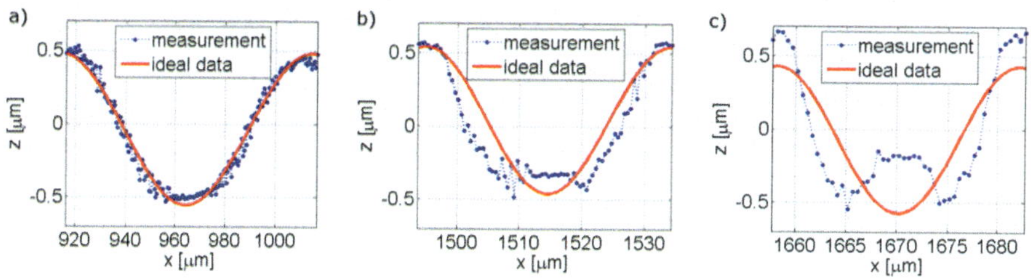

Fig. 3.3 A curved local surface can lead to artifacts in confocal microscopy [6]

explain the formation of artifacts in this phenomenon. However, no strategies were proposed for measuring unknown objects.

Another research group used a commercial confocal microscope to measure reference balls of varying diameters. The differences between the balls' measured and nominal profiles were used to train an artificial neural network (ANN) to correct errors during practical measurements [7]. However, this method is only valid when the detected object is ball-like, placed at the center of the field of view (FOV), and its dimensions fall within the range of the prepared reference balls. If the detected object is not ball-like, the authors compared the measurement results with a coherence scanning interferometer (CSI) to discuss validity by comparing the differences. In the later parts, we will discuss coherence scanning interferometry (CSI) and the similar problems it encounters when measuring tilted or curved surfaces.

Besides addressing the problem directly, some research approaches take different directions. For example, a specially designed confocal microscope can measure both surface position and slope [8]. As shown in Fig. 3.4, this system focuses different wavelengths on a single point but from different directions. Because surface slope can affect the direction of reflected light, the signal light is directed to different positions on an image sensor. Consequently, the position and slope of the detected point can be inferred. However, this method requires switching between many wavelengths to determine the local slope. Additionally, it can only measure slopes along one direction and is not easily modified to measure slopes in arbitrary directions.

Earlier proposed studies are even more intriguing. As shown in Fig. 3.5, a specially designed, rotational sector-shaped mask was used to "render" the MCM intensity signal. The term "render" means that the purpose was solely for better 3-D profile visualization. The rendering mechanism modulates the effective pupil size according to the direction of the reflected signal light. With this method, the grayscale visualization varies with the local slope of the detected object. Although only six levels of grayscale were demonstrated, this approach was significant in the era before the bloom of computational power [9]. Nevertheless, the critical point is that the strategy used in this research can quantify surface slope rather than just rendering for artistic or visualization purposes.

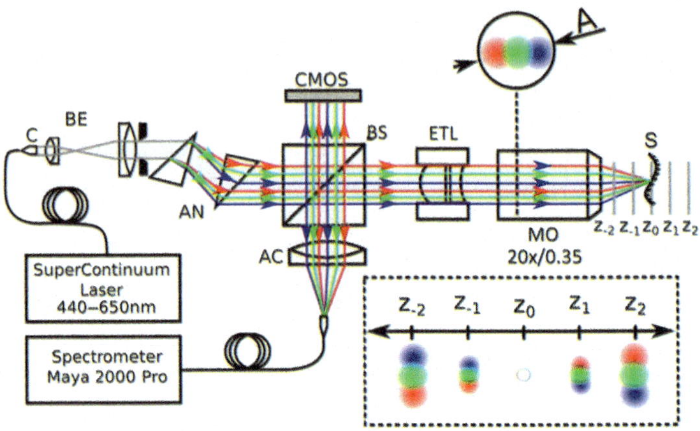

Fig. 3.4 A specially designed confocal system simultaneously measures the longitudinal position and local slope (only one direction) [8]

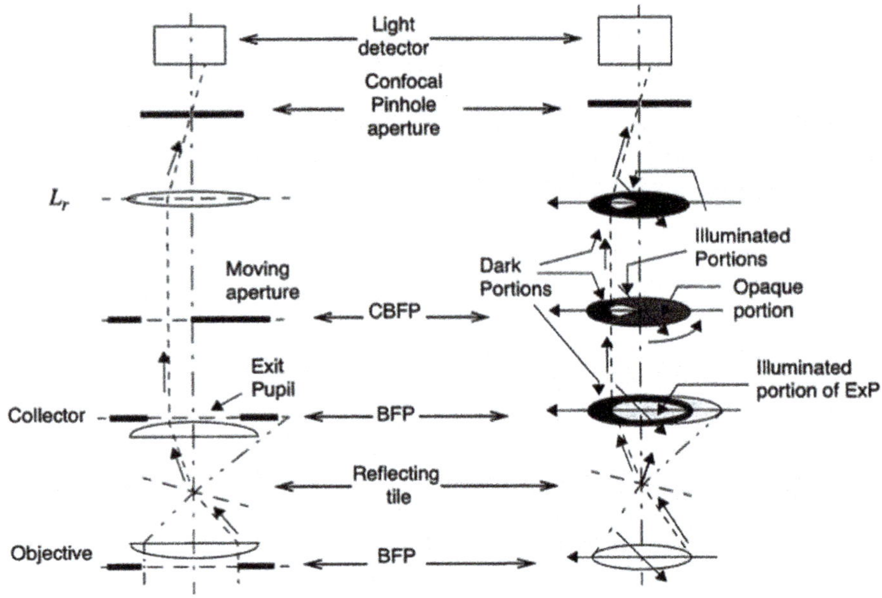

Fig. 3.5 A specially designed confocal system that can render the confocal signal intensity according to the local tilting of a detected object [9]

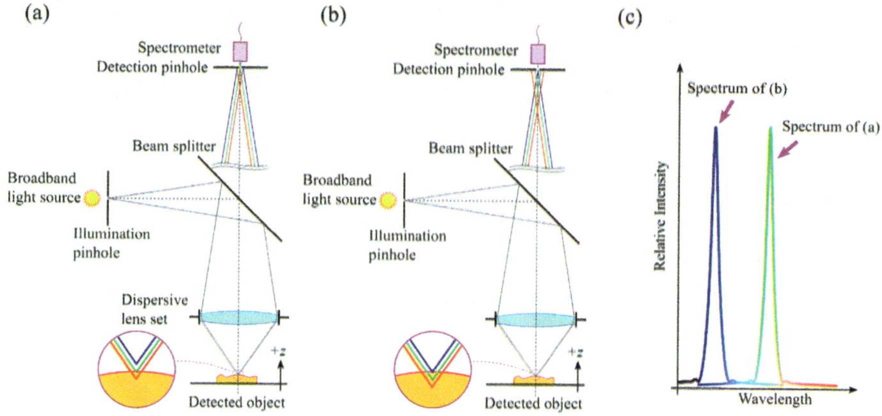

Fig. 3.6 Principle of chromatic confocal microscopy

3.3 Chromatic Confocal Microscopy

As a relative to MCM, chromatic confocal microscopy (CCM) uses a broadband light source and a dispersion component to focus different wavelengths to different positions, as shown in Fig. 3.6. Two spatial filters cooperate in sifting the most in-focus wavelength with the same principle as in MCM. The uniqueness of this method is the usage of a spectrometer instead of an intensity sensor to detect the spectrum of the light passing through the detection pinhole. The strongest wavelength component in the detected spectrum indicates that it just focuses on the object's surface. Knowing the property of the used dispersion component in advance allows reverse inference of the current surface position.

Similar phenomena can also be found in CCM. As shown in Fig. 3.7, it has been reported that the combination of system aberration and object surface slope can lead to significantly distorted spectra in CCM. The error is inevitable if these distorted spectra are used to determine the detected surface position [10].

3.4 Coherence Scanning Interferometry

CSI is another important optical technique used in surface profilometry. A typical CSI system using a Mirau objective is illustrated in Fig. 3.8, where a broadband light source provides the illumination. The beam splitter inside the Mirau objective separates the reference and measurement beams. As light propagates, the reference beam is reflected from the reference mirror, while the detected object reflects or scatters the measurement beam. These two beams merge and focus on the image sensor to form an interferogram. Due to the broad spectrum (low temporal coherence), the interfered signal is significant only within a concise longitudinal range, creating

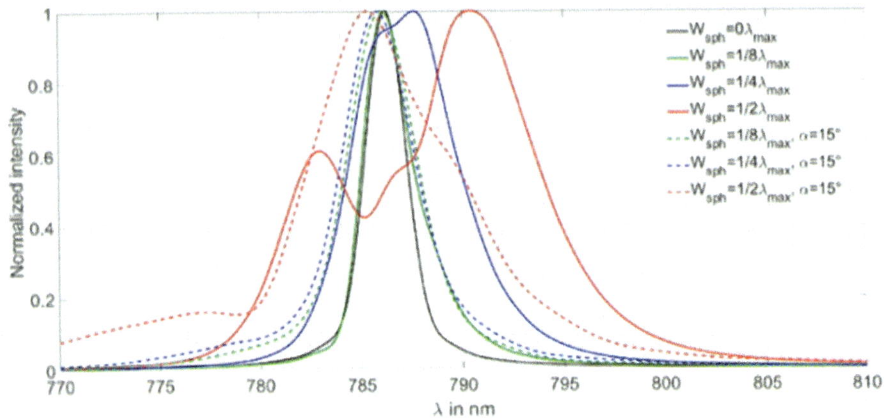

Fig. 3.7 Simulated CCM spectrum signals. The significantly distorted spectrums can lead to incorrect measurements [10]

narrow wave packets at each field point, as shown in the figure. Proper usage of the algorithm allows for extracting the peak position of the wave packet envelope, thereby measuring the surface height. The surface heights of all field points form the measured surface profile. With this principle, the longitudinal resolution of CSI is typically on the nanometer scale [11, 12].

However, despite its different mechanism, CSI also suffers from surface slope-dependent artifacts [13]. As reported, if the surface slope exceeds a specific limit, the interference fringes can become too dense, surpassing the sampling frequency of the image sensor. This results in the misinterpretation of the interferogram and causes artifacts. Similar phenomena have been demonstrated by measuring samples with sinusoidal profiles on several commercial CSI devices. The test results show that unwanted artifacts and phase jump counts are not rare. Furthermore, even in regions with small slopes, deviations from the ground truth still exist, leading to overestimation or underestimation of local curvature [14].

To address this issue, some researchers have developed a three-dimensional linear theory as an advanced model to describe the relationship between the detected object and the measured signal. This model assumes a linear invariant optical system, with the detected object expressed as a "foil model". Based on this, the three-dimensional point spread function (PSF) and modulation transfer function (MTF) of a real system can be measured experimentally and used to calibrate the optical system [15–19]. Unfortunately, this method cannot overcome the limitations caused by the insufficient sampling rate mentioned above. Nevertheless, it can reduce surface height measurement errors when a CSI system functions within its typical operational range.

Some methods have been developed based on statistical approaches. Random measurement noise can be eliminated by measuring random patches on a reference ball and averaging the results, leaving only systematic error [20]. System calibration can then be performed once the systematic error is known. This method is more

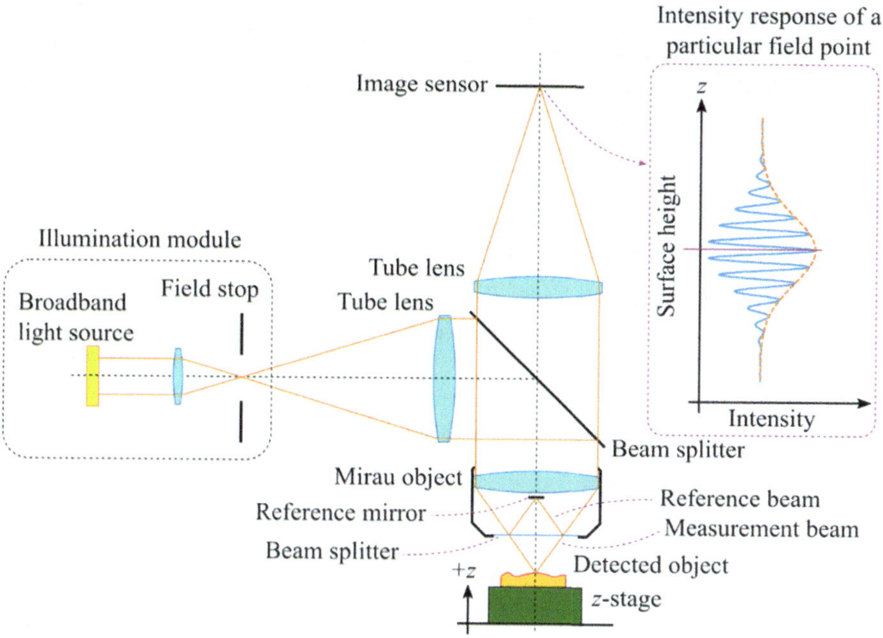

Fig. 3.8 A CSI system equipped with a Mirau objective

accessible and applicable to various optical measurement instruments. However, the calibration is only valid for measuring objects similar to the reference ball.

3.5 Surface-Dependent Error in Other Techniques

Other existing techniques suffer from similar problems. For example, in structured illumination microscopy (SIM) based profilometry, it has been pointed out that a tilted specular surface can cause transverse signal shifting, leading to a slope-dependent systematic error [21]. However, in widely used focus variation (FV) [22], instead of errors in surface height determination, the measured surface roughness may depend on the local slope [23, 24].

3.6 Intentionally Utilization of System Aberration

Although aberration should be minimal in most optical systems, some unique methods utilize aberration as the core of their measurement principle. For example, CCM uses chromatic aberration as its method kernel. Besides CCM, researchers also use optical systems that contain astigmatism to sense the position of an underlying object [25]. In this method, a four-quadrant photodiode is used as a photosensor. The detected object is placed under an optical system with astigmatism. Due to astigmatism, the light distribution on the four-quadrant photodiode forms an ellipsoid shape when the object is illuminated and imaged through the system. Consequently, the four-quadrant photodiode delivers four intensity signals with different values, sufficient for object positioning.

Similarly, a multiple-point version is proposed, and in this research, a pinhole array is used to provide multiple-point illumination, and an image sensor is used to detect multiple aberrated images. Similar to the previously mentioned research approach, the images are then transformed to height values to form the surface profile [26].

Although only objects with horizontal surfaces were measured in these two studies, it is not difficult to infer that the image may differ if an object is tilted. Furthermore, it suggests that height and tilt information can be restored from image-based data if the detected surface is observed point by point.

3.7 Potential Methods for Minimizing Surface-Dependent Errors

From the above review, it can be concluded that surface profile-dependent error is a common problem across the field of optical profilometry. Consequently, most researchers focus on modeling and explaining this error, utilizing the proposed models to characterize or calibrate real optical systems.

Reference balls are often required during system characterization or calibration. These reference objects minimize error if the subsequently measured object is of a similar shape and dimension. However, if the shape and dimensions differ, the validity of the characterization or calibration is degraded.

Additionally, some studies reveal the possibility of simultaneously measuring surface slope and height. These studies could offer a fundamental immunity to surface profile-dependent errors if they advance in the coming years.

References

1. Minsky M (1961) Microscopy apparatus US patent 3013467. USP Office, Ed. US
2. Wilson T et al (ed) (1990) Confocal microscopy. Academic Press, London
3. Webb RH (1996) Confocal optical microscopy. Rep Prog Phys 59(3):427
4. Wilson T (2011) Resolution and optical sectioning in the confocal microscope. J Microsc 244(2):113–121
5. Rahlves M, Roth B, Reithmeier E (2015) Systematic errors on curved microstructures caused by aberrations in confocal surface metrology. Opt Express 23(8):9640–9648

6. Mauch F et al (2012) Improved signal model for confocal sensors accounting for object depending artifacts. Opt Express 20(18):19936–19945
7. Béguelin J et al (2020) Correction of spherical surface measurements by confocal microscopy. Meas Sci Technol 31(7):075002
8. Pribošek J, Diaci J, Sinzinger S (2016) Aperture-coded confocal profilometry. Opt Lett 41(23):5523–5526
9. Conchello J-A et al (1997) Novel reflected light confocal profilometer. In: Three-dimensional microscopy: image acquisition and processing IV. International Society for Optics and Photonics, pp 101–112
10. Claus D, Nizami MR (2020) Influence of aberrations and roughness on the chromatic confocal signal based on experiments and wave-optical modeling. Surf Topogr: Metrol Prop 8(2):025031
11. De Groot P (2011) Coherence scanning interferometry. In: Optical measurement of surface topography. Springer, Berlin, Heidelberg, pp 187–208
12. De Groot P (2015) Principles of interference microscopy for the measurement of surface topography. Adv Opt Photonics 7(1):1–65
13. Lehmann P, Tereschenko S, Xie W (2016) Fundamental aspects of resolution and precision in vertical scanning white-light interferometry. Surf Topogr: Metrol Prop 4(2):024004
14. Gao F et al (2007) Surface measurement errors using commercial scanning white light interferometers. Meas Sci Technol 19(1):015303
15. Su R et al (2021) Scattering and three-dimensional imaging in surface topography measuring interference microscopy. JOSA A 38(2):A27–A42
16. Su R et al (2017) On tilt and curvature dependent errors and the calibration of coherence scanning interferometry. Opt Express 25(4):3297–3310
17. Su R et al (2019) High-accuracy surface measurement through modelling of the surface transfer function in interference microscopy. In: Applied optical metrology III. International Society for Optics and Photonics, p 1110205
18. Mandal R et al (2014) Coherence scanning interferometry: measurement and correction of three-dimensional transfer and point-spread characteristics. Appl Opt 53(8):1554–1563
19. Foreman MR et al (2013) Determination of the transfer function for optical surface topography measuring instruments—a review. Meas Sci Technol 24(5):052001
20. Zhou Y et al (2013) Application of the random ball test for calibrating slope-dependent errors in profilometry measurements. Appl Opt 52(24):5925–5931
21. Vogel M et al (2010) Tuning structured illumination microscopy (SIM) for the inspection of micro optical components. In: Proceedings of the DGaO, p A22
22. Nayar SK, Nakagawa Y (1994) Shape from focus. IEEE Trans Pattern Anal Mach Intell 16(8):824–831
23. Nikolaev N, Petzing J, Coupland J (2016) Focus variation microscope: linear theory and surface tilt sensitivity. Appl Opt 55(13):3555–3565
24. Hiersemenzel F et al (2012) Areal texture and angle measurements of tilted surfaces using focus variation methods
25. Fan K-C, Chu C-L, Mou J-I (2001) Development of a low-cost autofocusing probe for profile measurement. Meas Sci Technol 12(12):2137
26. Kang C-S et al (2010) One-shot measurement of surface profile using an astigmatic microscope system. In: Optical micro-and nanometrology III. International Society for Optics and Photonics, p 77180X

Chapter 4
Advanced Signal Processing for Diffractive Image Microscopy

4.1 Introduction

As a prerequisite for understanding the forthcoming chapters, relevant knowledge of signal processing techniques for diffractive image microscopy is presented, along with practical examples.

4.2 Normalized Cross-Correlation

Normalized cross-correlation (NCC) is a commonly used algorithm to quantify the similarity between two images or signals [1]. In image processing, it is typically defined by

$$s = \mathcal{S}[\mathbf{P}, \mathbf{Q}] = \frac{\sum_{m=1}^{M}\sum_{n=1}^{N} P_{m,n} Q_{m,n}}{\left[\left(\sum_{m=1}^{M}\sum_{n=1}^{N} P_{m,n}^2\right)\left(\sum_{m=1}^{M}\sum_{n=1}^{N} Q_{m,n}^2\right)\right]^{1/2}} \tag{4.1}$$

where **P** and **Q** are two images represented by two matrices of the same size $M \times N$. And $P_{m,n}$, $Q_{m,n}$ denotes the (m, n) entries in **P** and **Q** respectively. $\mathcal{S}(\mathbf{P}, \mathbf{Q})$ means an operator $\mathcal{S}$ acts on **P** and **Q** acquires the similarity value (or the NCC value) s. With this definition, the value s can only be within [0, 1], where 0 denotes totally unlike, and 1 denotes perfectly similar. Figure 4.1 shows an example that three images compared by NCC mutually; the highest value of 0.74 between the spiral and the circle means they are the most similar images in these three, agreeing with human vision's intuition.

Besides the above definition, there is another similar definition called zero mean NCC (ZNCC), which is written as

L. Chen et al., *Diffractive Image Microscopy for 3D Imaging*, Springer Tracts in Mechanical Engineering, https://doi.org/10.1007/978-981-97-7782-2_4

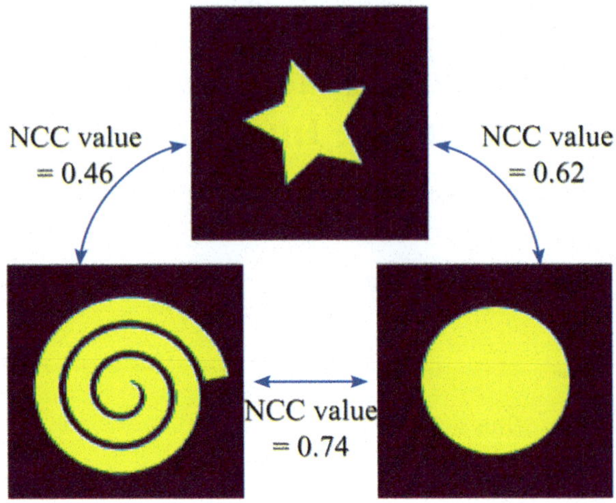

Fig. 4.1 NCC values between three different images

$$s' = \mathcal{S}'[\mathbf{p}, \mathbf{q}] = \frac{\sum_{m=1}^{M} \left[(p_m - \overline{p})(q_m - \overline{q})\right]_m}{\left[\left(\sum_{m=1}^{M} (p_m - \overline{p})^2\right)\left(\sum_{m=1}^{M} (q_m - \overline{q})^2\right)\right]^{1/2}} \quad (4.2)$$

where **p** and **q** are two vectors. p_m and q_m are the components of **p** and **q** respectively. $\overline{p}$ and $\overline{q}$ are the mean value of **p** and **q**, which are defined by

$$\begin{cases} \overline{p} = \frac{1}{M} \sum\limits_{m-1}^{M} p_m \\ \overline{q} = \frac{1}{M} \sum\limits_{m-1}^{M} q_m \end{cases} \quad (4.3)$$

The main difference between NCC and ZNCC is not the dimension but the subtraction of the mean. It makes no difference mathematically because an image can be unraveled to a long 1-D vector. However, ZNCC has a more extensive range s' $[-1, 1]$ because the mean subtraction results in higher sensitivity than NCC. Nevertheless, this does not mean it is always better.

NCC is independent of a dark background with zero or extremely low grayscale values [they have no contribution in Eq. (4.1)]. This means that the NCC value is unaffected by the size of the cropped ROI if the signal is only at the center and the background is dark [3]. However, if the items being compared are not images but vectors containing physical quantities of interest, ZNCC may be better due to its higher sensitivity.

Another important consideration is that when comparing specific physical quantities, the feature normalization process is crucial to ensure all physical quantities have a similar value range and zero mean. Without normalization, the NCC or ZNCC value can be very insensitive to quantities with smaller values. This insensitivity does not imply that these quantities are unimportant; it may simply be due to a larger unit being used. The feature normalization is done by

$$q'_{\text{phy}} = \frac{q_{\text{phy}} - \text{EXP}(q_{\text{phy}})}{\text{STD}(q_{\text{phy}})} \tag{4.4}$$

where q_{phy} is the original physical quantity and q'_{phy} is the feature normalized quantity. EXP and STD mean to take the expectation and standard deviation, respectively. Frequently, a quantity with its original scale and mean needs to be recovered from a normalized quantity. This is given by

$$q_{\text{phy}} = q'_{\text{phy}}\text{STD}(q_{\text{phy}}) + \text{EXP}(q_{\text{phy}}) \tag{4.5}$$

With NCC, ZNCC, and feature normalization, the similarity between experimental images or their extracted features can be quantitatively analyzed.

4.3 Levenberg–Marquardt Algorithm

A common scenario involves fitting a set of experimental data with a designed equation. The fitted equation can then be used to analytically represent the experimental data for interpolation, system characterization, or feature extraction.

Now, suppose an experiment is conducted, and a set of experimental data is collected. It consists of a collection of $(\mathbf{x}_j, y_j)$ with $j = 1 \sim N_{\text{data}}$, where $\mathbf{x}_j$ is an input vector containing N_{var} values and y_j is the corresponding output value. In this case, a designed function can fit the data by minimizing the sum of squared error (SSE) between the predicted and true values.

where

$$f_{\text{fit}}(\mathbf{x}; \mathbf{p}) : \mathbb{R}^{N_{\text{var}}} \rightarrow \mathbb{R}^1 \tag{4.6}$$

$$\mathbf{x}_j = \begin{bmatrix} x_{j,1} & \cdots & x_{j,N_{\text{var}}} \end{bmatrix}^T \tag{4.7}$$

is the input vector, and

$$\mathbf{p} = \begin{bmatrix} p_1 & \cdots & p_{N_{\text{para}}} \end{bmatrix}^T \tag{4.8}$$

is the parameter vector $\mathbf{p}$ that contains N_{para} coefficients, which determine the shape of f_{fit}. The SSE is written as

$$E_{ss}(\mathbf{p}) = \mathbf{r}^T\mathbf{r} \tag{4.9}$$

with

$$\mathbf{r} = \left[f_{\text{fit}}(\mathbf{x}_j; \mathbf{p}) - y_j\right]^T, \ j = 1 \sim N_{\text{data}} \tag{4.10}$$

If the mutual relationship between $\mathbf{x}_j$ and y_j is simple enough, conventional linear regression is sufficient to represent the underlying relationship behind the data. However, if the mutual relationship between $\mathbf{x}_j$ and y_j is rather complex, a nonlinear equation may be necessary to fit them properly. In this case, the Levenberg–Marquardt algorithm (LMA) is one of the most robust methods [4, 4, 5] to minimize the SSE. The LMA is an adaptive algorithm that iteratively updates the parameters, **p**, according to the strategy.

$$\mathbf{p}_{k+1} = \mathbf{p}_k - \left(\mathbf{J}_r^T \mathbf{J}_r + \mu_k \mathbf{I}\right)^{-1} \mathbf{J}_r^T \mathbf{r} \tag{4.11}$$

where $\mathbf{p}_k$ and $\mathbf{p}_{k+1}$ are the current and next parameter vectors, respectively. $\mathbf{J}_r$ is the Jacobian matrix of **r** with respect to **p**, which is written explicitly as

$$\mathbf{J}_r = \begin{bmatrix} \frac{\partial r_1}{\partial p_1} & \cdots & \frac{\partial r_1}{\partial p_{N_{\text{para}}}} \\ \vdots & \ddots & \vdots \\ \frac{\partial r_{N_{\text{data}}}}{\partial p_1} & \cdots & \frac{\partial r_{N_{\text{data}}}}{\partial p_{N_{\text{para}}}} \end{bmatrix} \tag{4.12}$$

where both $\mathbf{J}_r$ and **r** take $\mathbf{p}_k$ as the argument. **I** is the identity matrix and μ_k is a positive coefficient that is adjusted according to the real descent. The primary principle to adjust μ_k is gradually increasing it if the SSE increases. As μ_k gradually increases, Eq. (4.11) gradually approaches

$$\mathbf{p}_{k+1} = \mathbf{p}_k - \frac{1}{\mu_k} \mathbf{J}_r^T \mathbf{r}. \tag{4.13}$$

This is the parameter update strategy of the gradient descent method. Therefore, the decrease of the SSE can be ensured. On the contrary, if the SSE drops smoothly, μ_k is gradually decreased so that Eq. (4.11) gradually approaches

$$\mathbf{p}_{k+1} = \mathbf{p}_k - \left(\mathbf{J}_r^T \mathbf{J}_r\right)^{-1} \mathbf{J}_r^T \mathbf{r} \tag{4.14}$$

In this case, the algorithm is an approximated Gauss–Newton algorithm, which uses the stationary point of an approximate quadratic function to update the parameter, leading to faster optimization. Fitting experiment data with complex nonlinear equations is achievable with LMA and a set of reasonable initial parameters.

For example, Fig. 4.2 shows a fitting example with LMA, where the noisy data in Fig. 4.2a is generated by

$$y(x; p_1, p_2, p_3, p_4) = \sin(p_1 \exp(p_2 x)) \exp(-p_3 (x - p_4)^2) + y_{\text{noise}} \tag{4.15}$$

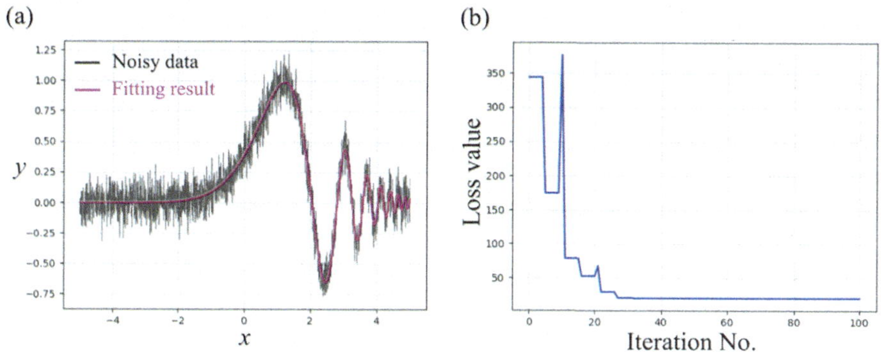

Fig. 4.2 **a** Original noisy data and the LMA fitting result. **b** The loss value changes during the optimization process

where the parameters $p_1 \sim p_4$ are 0.5, 0.9, 0.2, and 1.0 respectively. The last term y_{noise} is a normally distributed random variable with a standard deviation 0.1. The fitting starts with all parameters to be 1.0, resulting in a good representation shown in Fig. 4.2a. The SSE and iteration number relationship is plotted in Fig. 4.2b. In which the spike says that the SSE increases suddenly, but the algorithm adaptive switches to the gradient descent-similar mode, and then the optimization goes on the rail again.

4.4 Multiple Layer Perception

Multiple layer perceptron (MLP) is one of the most basic types of ANN [10]. In which several layers are connected in series, each layer contains multiple neurons. Referring to Fig. 4.3a, a neuron contains two components, an activation function, f_a, and a bias term b [8]. The activation function provides an output value to the downstream neurons according to

$$v = f_a(S_w + b) \tag{4.16}$$

where v is the output value, S_w is the summation of weighted output values from the upstream neurons. Now consider there are two cascaded layers as shown in Fig. 4.3b, in which layer k has N_k neurons and layer $k+1$ has N_k neurons. In this case, the connection between these two layers can be expressed by a $N_{k+1} \times N_k$ matrix $\mathbf{W}_k = [w_{k,pq}]$, in which $w_{k,pq}$ denotes the weighting for the output of the p-th neuron in layer k when transmitted to the q-th neuron in the layer $k+1$. With an assumption that all neurons have the same activation function, the output values of the layer $k+1$ can be expressed as a vector $\mathbf{v}_{k+1}$, which is given by

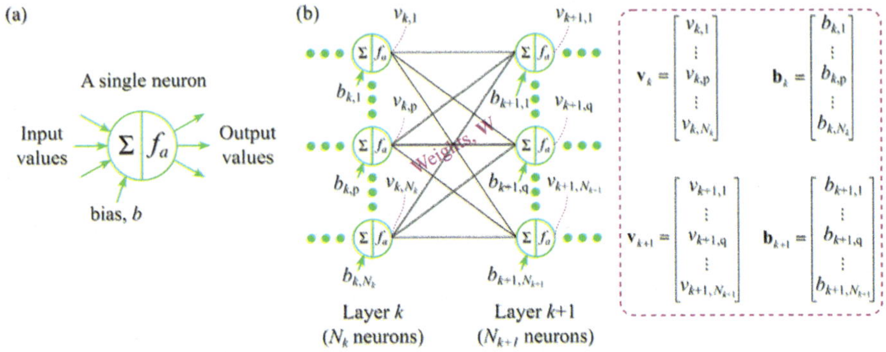

Fig. 4.3 **a** A single neuron and **b** two cascaded neuron layers

$$\mathbf{v}_{k+1} = f_a(\mathbf{W}_k \mathbf{v}_k + \mathbf{b}_{k+1}). \tag{4.17}$$

where $\mathbf{v}_k$ is a vector that contains the output of layer k. $\mathbf{b}_{k+1}$ is a vector that contains the biases of the layer $k+1$.

If K layers are connected in series to form an MLP, then the output vector of the final layer can be written as

$$\mathbf{v}_K = f_a(\mathbf{W}_{K-1}(f_a(\mathbf{W}_{K-2} f_a(\mathbf{W}_{K-3}(\cdots) + \mathbf{b}_{K-2}) + \mathbf{b}_{K-1})) + \mathbf{b}_K) \tag{4.18}$$

where $\mathbf{v}_k$ is the output vector of the last layer, resulting from a function composition. It is important to note that typically, the output vector of the first layer is simply the input vector of the entire MLP, which corresponds to the inputs of the experimental data for fitting purposes [12]. Additionally, the last layer does not have activation functions. However, this detail is ignored here for simplicity.

As a result, the overall function of the whole MLP expressed by Eq. (4.18) provides a vast solution space, enabling the MLP to simulate highly complex functions. This means that even if we do not know how to design an analytical equation to fit a certain problem, a suitable analytical equation is highly likely to be contained within the solution space provided by such an MLP.

The next step is similar to conventional fitting algorithms: the partial derivatives of the loss value with respect to all weights and biases are required for gradient descent-based algorithms to optimize the weights and biases within the MLP. The commonly used loss value is the SSE (sum of squared errors) for regression. According to the experimental data, the SSE refers to the difference between the final output vector, $\mathbf{v}_k$, and the target output vector.

Due to the cascaded structure of the MLP, a high-efficiency algorithm called backpropagation [2] is commonly applied to calculate all the partial derivatives from the last layer to the first layer. Once the partial derivative of a particular weight or bias is computed, it is updated according to gradient descent-based algorithms. One popular and robust algorithm is the adaptive moment estimation (Adam) [9], which is used in the context of this book.

With MLP, complex experimental data can be fit without explicitly designing an analytical equation. This is particularly important when the problem at hand is overwhelmingly complex. For example, the generated data in Fig. 4.2a can also be fitted by an MLP, as shown in Fig. 4.4. The fitting result and the loss value are shown in Fig. 4.5, where the adopted activation function is the logistic function defined by [12]

$$f_{\mathrm{a}}(x) = \frac{1}{1 + e^{-x}} \tag{4.19}$$

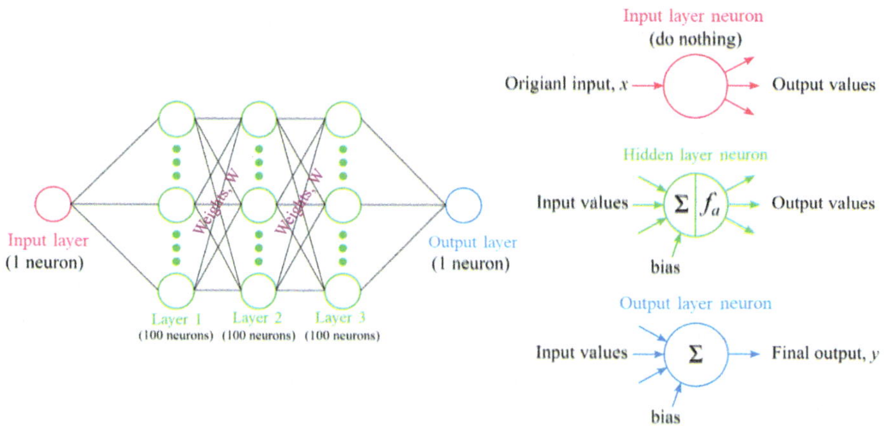

Fig. 4.4 A three-layer MLP used to fit the same generated data used in Fig. 4.2

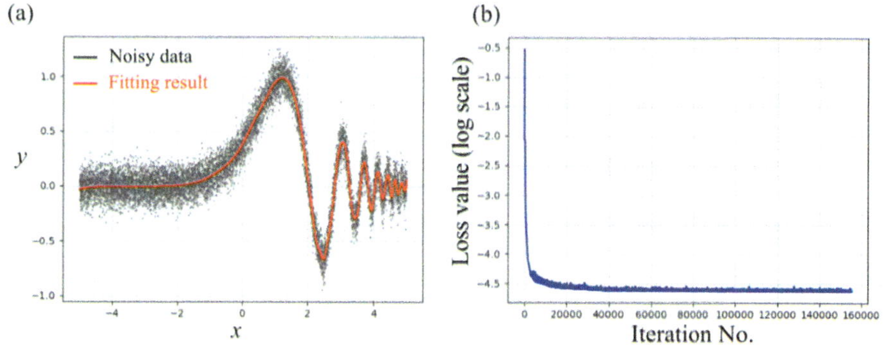

Fig. 4.5 **a** Fitting result of a three hidden layer MLP and **b** the loss value versus iteration number (epochs)

4.5 Deep Learning

4.5.1 Encoder and Decoder Deep Learning Model

Deep learning has been widely applied to the problem of image deblurring with great success. Image deblurring removes blur from an image, which can be caused by camera shake, motion, or defocus blur.

Deep learning models for image deblurring typically use convolutional neural networks (CNNs) to learn the transformation from blurry images to clear images. These models can be trained on large datasets of blurred and sharp image pairs to learn how to effectively remove various types of blur.

Deep learning models for image deblurring typically use convolutional neural networks (CNNs) to acquire knowledge of a transformation from blurry images to clear images. These models can be trained on large datasets of blurred and sharp image pairs to learn how to remove different types of blur.

One common approach is to use an encoder-decoder architecture (Fig. 4.6), where the encoder processes the blurred image to create a fixed-length representation or "context vector", and the decoder generates a sharp, deblurred image from this context vector. Other approaches include using generative adversarial networks (GANs) or incorporating additional components such as skip connections or attention mechanisms (Fig. 4.7).

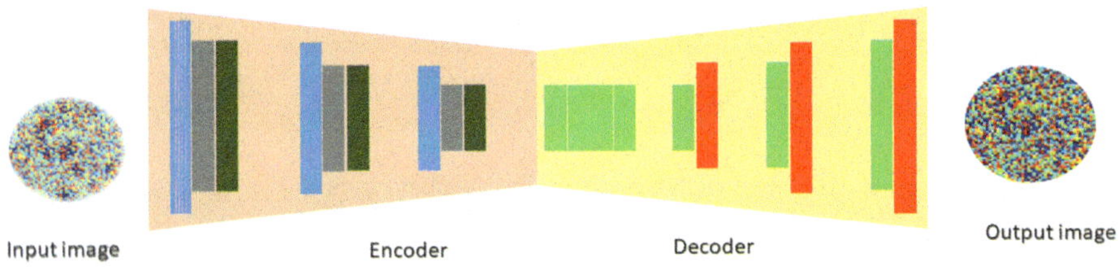

Fig. 4.6 Architecture of the encoder and decoder

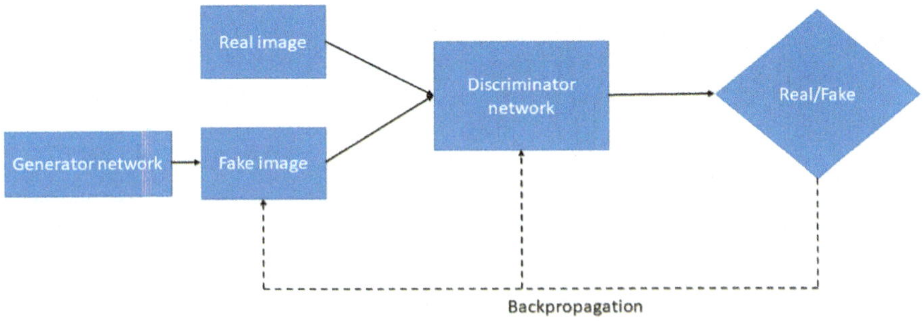

Fig. 4.7 A schematic diagram of GANs architecture

Deep learning-based image deblurring has many practical applications, including improving the quality of photographs taken with handheld cameras or smartphones, restoring old or damaged photographs, and enhancing the resolution of medical or satellite images.

An encoder-decoder model is a popular deep learning architecture that is frequently used for natural language processing and sequence-to-sequence tasks. The model comprises two primary components: an encoder and a decoder.

The encoder is responsible for taking in an input sequence and processing it to create a fixed-length representation, also known as a "context vector". This vector takes the pivotal information from the input sequence and is subsequently fed into the decoder. The decoder is responsible for generating the output sequence based on the information contained in the context vector. One way the decoder can be trained is to generate the output sequence one element at a time, using the previously generated elements as additional input.

Encoder-decoder models are successfully applied to a broad range of activities, such as machine translation, text summarization, and image captioning. They are powerful tools for handling sequential data and can be trained holistically using supervised learning techniques.

The encoder and decoder make it possible to train these tasks using supervised learning techniques that involve end-to-end training. The training data consists of pairs of sharp images and blurry images. The model is trained to minimize the difference between the deblurred image generated by the decoder and the corresponding sharp image in the training data.

Encoder-decoder models have shown promising results for image deblurring. They can be used to restore sharpness to images degraded by motion, defocus, or other types of blur.

One additional thing to note is that many variations of the encoder-decoder architecture can be used for image deblurring. For example, some models use convolutional neural networks (CNNs) for the encoder and decoder, while others use recurrent neural networks (RNNs) or a combination of both (Fig. 4.8).

Additionally, some models incorporate skip connections or attention mechanisms to enhance performance. The choice of architecture and components depends on the specific characteristics of the deblurring task and the data being used.

Overall, encoder-decoder models provide a flexible and powerful framework for image deblurring and can be adapted to a wide range of scenarios.

4.5.2 Loss in the Deep Learning Model

The primary goal of deep learning is to minimize the difference between the model's predictions and the actual outcomes. A loss function quantifies this difference, enabling the optimization of model parameters. By minimizing the error, the model improves its predictive accuracy. The choice of loss function depends on the specific

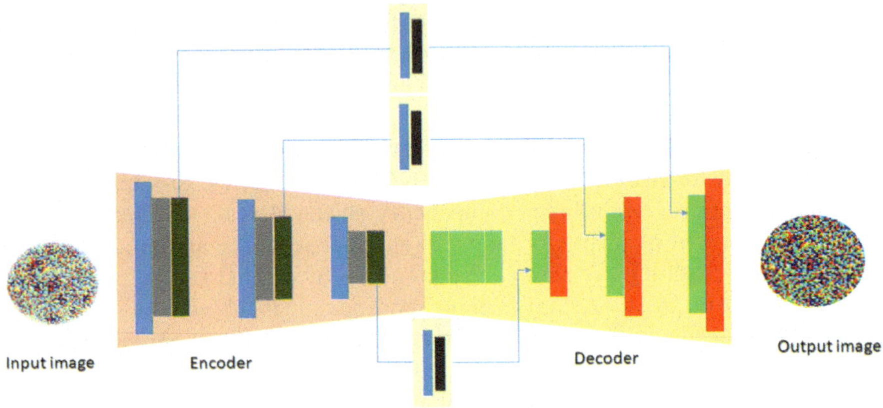

Fig. 4.8 Schematic diagram of the encoder and decoder with RNN architecture

problem being addressed. Several loss functions are commonly used in deep learning, some described here.

1. Mean Squared Error (MSE): MSE is primarily used in regression problems. It calculates the average of the squared differences between the ground truth and the predicted output.
2. Binary Cross-Entropy (BCE): This loss function is used for binary classification problems. It computes the cross-entropy between the target (ground truth) output and the predicted output. The formula for BCE is:

$$\text{BCE} = -\left(y_{\text{actual}} \cdot \log\left(y_{\text{pred}}\right) + (1 - y_{\text{actual}}) \cdot \log\left(1 - y_{\text{pred}}\right)\right) \tag{4.20}$$

   where y_{pred} is the predicted output and y_{actual} is the actual output (0 or 1).
3. Categorical Cross-Entropy (CCE): This loss function is used for multi-class classification problems. It calculates the cross-entropy between the predicted output and the ground truth, where the actual output is represented as a one-hot encoded vector. The formula for CCE is:

$$\text{CCE} = -\sum \left(y_{\text{actual}} \cdot \log\left(y_{\text{pred}}\right)\right) \tag{4.21}$$

   where y_{pred} is the predicted output and y_{actual} is the actual output.
4. Huber Loss: This loss function is used in regression problems, similar to the MSE loss function. However, unlike the MSE, Huber loss is less sensitive to outliers in the data, making it a more robust loss function.

   The formula for Huber loss is expressed as:

$$L = \frac{1}{n} \cdot \sum L_i \tag{4.22}$$

where n is the number of samples, L_i is the loss for the i_{th} sample, and the loss for the i_{th} sample is calculated as:

$$
\begin{aligned}
L_i &= 0.5 \cdot \left(y_{\text{pred}} - y_{\text{actual}}\right)^2 \quad \text{if } \left|y_{\text{pred}} - y_{\text{actual}}\right| \leq \text{delta} \\
L_i &= \text{delta} \cdot \left(y_{\text{pred}} - y_{\text{actual}}\right) - 0.5 \cdot \text{delta}^2 \quad \text{if } \left|y_{\text{pred}} - y_{\text{actual}}\right| > \text{delta}
\end{aligned} \tag{4.23}
$$

where y_{pred} is the predicted output and y_{actual} is the actual output.

Delta is a hyperparameter that regulates the threshold, determining when the loss function transitions from quadratic to linear. Suppose the absolute difference between the ground truth output and the actual output is smaller than or equal to the delta value. In that case, the Huber loss behaves similarly to the mean squared error (MSE) loss, resulting in a quadratic loss function.

The advantage of Huber loss over MSE loss is its robustness to outliers in the data. This robustness arises because the quadratic component of the Huber loss function is less sensitive to large errors than the pure quadratic nature of the MSE loss. However, it remains sensitive enough to minor errors, ensuring the model's accuracy for most data points. Therefore, Huber loss is a good choice when the data contains outliers or the model needs to be robust to noise in the data.

Although the aforementioned loss functions are widely used in deep learning, it is important to note that various other loss functions exist. The selection of a loss function depends on the nature of the problem being addressed and the type of output the model generates. Therefore, choosing the appropriate loss function is crucial to ensure effective model training and accurate results.

4.5.3 Activation Function in Deep Learning

Activation functions play a vital role in introducing nonlinearity into deep learning models. Without nonlinearity, the model would be limited to linear relationships, making it unsuitable for complex tasks like speech or image recognition. Several activation functions are commonly used in deep learning, and here are some of the most popular ones:

1. Sigmoid: The sigmoid function is a commonly used activation function in neural networks. It maps any input value to a positive fractional value between 0 and 1. The sigmoid function is defined as:

$$f(x) = \frac{1}{(1 + \exp(-x))} \tag{4.24}$$

   where x is the input to the function.
2. ReLU: The ReLU (Rectified Linear Unit) function is another extensively used activation function in deep learning. It is defined as:

$$f(x) = \max(0, x) \tag{4.25}$$

where x is the function input. The ReLU function is simple and computationally efficient, making it popular in deep learning.

3. Softmax: The Softmax function is frequently used as the activation function in the output layer of a neural network for classification problems. It maps a vector of input values to a probability distribution over the output classes. The Softmax function is defined as:

$$f(x_i) = \frac{\exp(x_i)}{\sum \exp(x_i)} \tag{4.26}$$

where x_i is the input to the i_{th} neuron in the output layer.

The choice of activation function depends on the nature of the problem and the specific architecture of the neural network. Some activation functions are better suited for specific problems or architectures, while others may be less effective.

References

1. Yoo J-C, Han TH (2009) Fast normalized cross-correlation. Circ Syst Signal Process 28(6):819–843
2. Wu G-W, Nguyen DT, Chen L-C (2019) Novel full-field micro surface profilometry using diffractive image correlation without vertical scanning. Opt Lett 44(14):3534–3537
3. Marquardt DW (1963) An algorithm for least-squares estimation of nonlinear parameters. J Soc Ind Appl Math 11(2):431–441
4. Roweis S (1996) Levenberg-marquardt optimization. Notes, University of Toronto
5. Moré JJ (1978) The Levenberg-Marquardt algorithm: implementation and theory. In: Numerical analysis. Springer, Berlin, Heidelberg, pp 105–116
6. Gavin HP (2019) The Levenberg-Marquardt algorithm for nonlinear least squares curve-fitting problems. Department of Civil and Environmental Engineering, Duke University, 1–19
7. Sarle WS (1994) Neural networks and statistical models
8. Wu G-W, Chen L-C (2023) Resolving measurement ambiguity in diffractive image microscopy for 6DOF surface measurement using designed aberration and multiple-layer perceptron. Int J Adv Manuf Technol. https://doi.org/10.1007/s00170-023-11147-0
9. Wu G-W (2022) Theory and implementation of microscopic diffractive image profilometry. Ph.D. thesis, National Taiwan University
10. Hecht-Nielsen R (1992) Theory of the backpropagation neural network. In: Neural networks for perception. Academic Press, pp 65–93
11. Kingma (2014) Adam: A method for stochastic optimization. arXiv
12. Wu G-W, Chen L-C (2021) Precise 3-D microscopic profilometry using diffractive image microscopy and artificial neural network in single-exposure manner. Opt Lasers Eng 147:106732

Chapter 5
Formation of Diffractive Images

5.1 Introduction

This chapter proposes a physical model to describe the skeleton of a diffractive image microscope. The proposed model simulates signal images by placing a tilted flat mirror (TFM) under a virtual system. It also allows for studying the influence of aberration and surface roughness. These simulated images are useful for data processing and inverse problem testing.

5.2 Model Derivation

Considering the optical system depicted in Fig. 5.1, an ideal point source (IPS) serves as the light source, and a tilted flat mirror (TFM) is detected. The IPS is assumed to be monochromatic and coherent. After thoroughly studying the monochromatic and coherent case, the model can be extended to broadband and incoherent sources by summing up the final intensity signal on the image sensor for multiple wavelengths. A convolution operation on the image sensor plane can also simulate the case with a finite-size light source [1].

First, the light emitted from the IPS is directed downward by the beam splitter and then focused toward point o by the optical system. Note that the "optical system" is represented by a two-dimensional pupil function that exists only in the pupil plane, i.e., the plane of $z = d_{\mathrm{po}}$. Now, suppose the optical system is aberration-free. In that case, the wavefronts passing downward through the pupil plane remain spherical wavefronts centered on point o. As illustrated in Fig. 5.2, a spherical wavefront, w_{f}, converging to a point P_{converge} can be expressed as:

$$w_{\mathrm{f}} = \exp\left[-ik\left(\sqrt{d_{\mathrm{converge}}^2 + r^2} - d_{\mathrm{converge}}\right)\right], \tag{5.1}$$

L. Chen et al., *Diffractive Image Microscopy for 3D Imaging*, Springer Tracts in Mechanical Engineering, https://doi.org/10.1007/978-981-97-7782-2_5

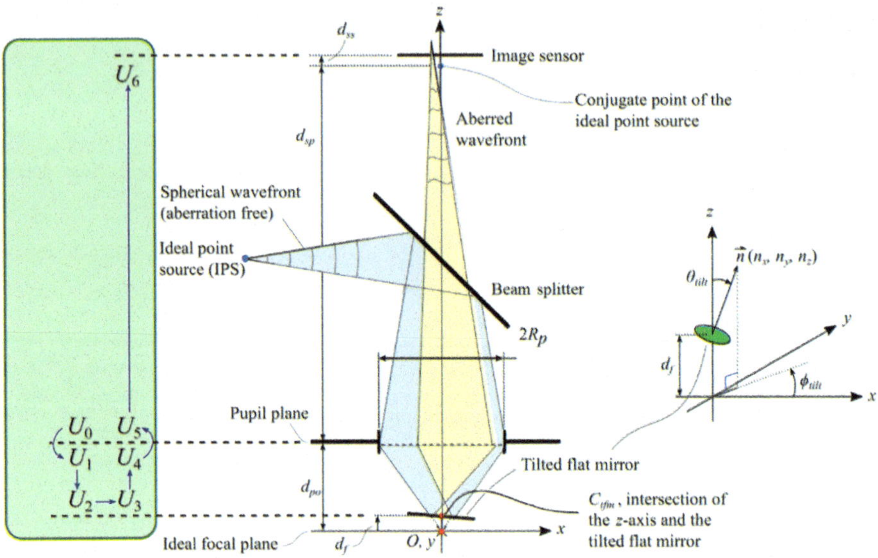

Fig. 5.1 System configuration used for model derivation

where k is the wavenumber defined by $k = 2\pi/\lambda$; λ is the wavelength of light. i is the imaginary unit; d_{converge} is the distance from the center of the wavefront to the point P_{converge}; and r is the radial distance from the optical axis that passes through the center of the wavefront and point P_{converge}.

With Eq. (5.1), the wavefront passes downward through the pupil plane and can be directly written as

$$U_{\text{ub}} = \exp\left[-ik\left(\sqrt{d_{\text{po}}^2 + r^2} - d_{\text{po}}\right)\right], \tag{5.2}$$

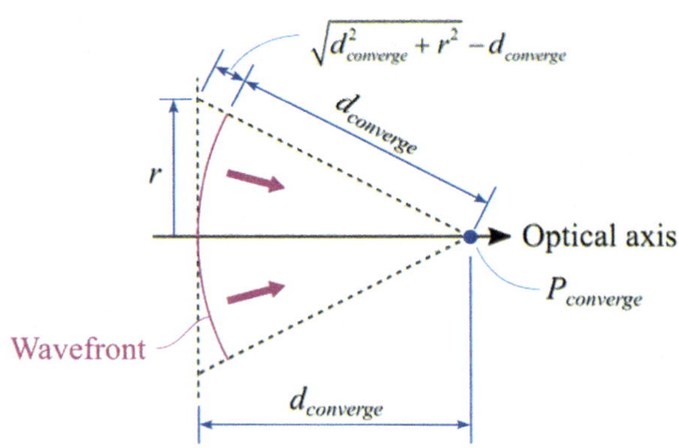

Fig. 5.2 Spherical wavefront geometry

where U_{ub} is an unbounded light field, d_{po} is the distance between the pupil plane and the ideal focal plane ($x - y$ plane) and $r = \sqrt{x^2 + y^2}$ is the radial distance from the optical axis to arbitrary points on the pupil plane. If the optical system is not aberration-free, the wavefront expressed by Eq. (5.2) should be multiplied by an extra term that describes the system aberration as

$$T_{\text{aberr}} = T_{\text{aberr}}(x, y). \tag{5.3}$$

Meanwhile, the optical system should have a finite pupil size with a given radius R_p, represented by the function:

$$\text{circ}\left(\frac{r}{R_p}\right) = \begin{cases} 1, & r \le R_p \\ 0, & r > R_p \end{cases}. \tag{5.4}$$

That limits the region through which light can pass through. Therefore, the complete expression of the wavefront that passes downward through the pupil plane is denoted U_1 and can be written as

$$U_1(x, y) = \exp\left(-ik\left(\sqrt{d_{\text{po}}^2 + r^2} - d_{\text{po}}\right)\right) T_{\text{aberr}}(x, y) \text{circ}\left(\frac{r}{R_p}\right). \tag{5.5}$$

Next, the light field on a horizontal plane that passes through the point C_{tfm} is desired (i.e., the plane of $z = d_{\text{f}}$). This light field is denoted as U_2, and it can be calculated by applying the Rayleigh-Sommerfeld diffraction integral (RSDI), which gives that

$$U_2(x, y) = \frac{1}{i\lambda} \int_{-\infty}^{\infty} \int_{-\infty}^{\infty} U_1(\xi, \eta) \exp(ikr_{\text{eculi}}) \frac{\Delta z}{r_{\text{eculi}}^2} \mathrm{d}\xi \mathrm{d}\eta. \tag{5.6}$$

where $\Delta z = d_{\text{po}} - d_{\text{f}}$, and

$$r_{\text{eucli}} = \sqrt{(x - \xi)^2 + (y - \eta)^2 + \Delta z^2}. \tag{5.7}$$

As shown in Fig. 5.1, the position and the orientation of the flat mirror can be defined by d_{f} and its unit normal vector (n_x, n_y, n_z), or defined by d_{f}, θ_{tilt}, and ϕ_{tilt}. The parameters such as d_{f}, θ_{tilt}, and ϕ_{tilt} are referred to as the posture parameters in the remainder of the content. The choice between these two expressions depends on convenience. These two expressions can be converted mutually according to

$$\begin{cases} n_x = \sin\theta_{\text{tilt}} \cos\phi_{\text{tilt}} \\ n_y = \sin\theta_{\text{tilt}} \sin\phi_{\text{tilt}} \end{cases} \tag{5.8}$$

and

$$\begin{cases} \theta_{\text{tilt}} = \cos^{-1}(n_z) \\ \phi_{\text{tilt}} = \tan^{-1}\left(n_y/n_x\right) \end{cases}. \tag{5.9}$$

With the unit normal vector (n_x, n_y, n_z), the surface of the TFM can be expressed as

$$z = -\frac{n_x}{n_z}x - \frac{n_y}{n_z}y + d_{\text{f}}. \tag{5.10}$$

Therefore, if the TFM reflected an incoming light wave, it can induce an optical path difference (OPD) described by

$$-2\left[\left(-\frac{n_x}{n_z}x - \frac{n_y}{n_z}y + d_{\text{f}}\right) - d_{\text{f}}\right] = 2\left(\frac{n_x}{n_z}x + \frac{n_y}{n_z}y\right) \tag{5.11}$$

with respect to the point $C_{\text{tfm}}(0, 0, d_{\text{f}})$, where the "– 2" in Eq. (5.11) is due to the reflection. As a consequence, the transmission function of the flat mirror is written as:

$$\exp\left[ik2\left(\frac{n_x}{n_z}x + \frac{n_y}{n_z}y\right)\right]. \tag{5.12}$$

The light field, after being reflected by the TFM, is expressed as:

$$U_3 = U_2 \exp\left[ik2\left(\frac{n_x}{n_z}x + \frac{n_y}{n_z}y\right)\right]. \tag{5.13}$$

Similarly, when the reflected wavefront reaches the pupil plane, it can be calculated according to U_3 by using the RSDI again, denoted as U_4 and written as:

$$U_4(x, y) = \frac{1}{i\lambda}\int_{-\infty}^{\infty}\int_{-\infty}^{\infty} U_3(\xi, \eta)\exp(ikr_{\text{eculi}})\frac{\Delta z}{r_{\text{eculi}}^2}\text{d}\xi\,\text{d}\eta. \tag{5.14}$$

After U_4 passing upward through the pupil plane, it can be multiplied by the transmission function again to form

$$\exp\left(-ik\left(\sqrt{d_{\text{po}}^2 + r^2} - d_{\text{po}}\right)\right)\exp\left(-ik(\sqrt{d_{\text{sp}}^2 + r^2} - d_{\text{sp}})\right)T_{\text{aberr}}(x, y). \tag{5.15}$$

In Eq. (5.15), the first exponential term refers to transforming the diverging wavefronts reflected from the flat mirror into planar wavefronts. The second exponential term refers to transforming these planar wavefronts into converging wavefronts toward the conjugate point of the IPS. The last term represents the influence of system aberration, as explained earlier. Therefore, the light field passing upward through the pupil plane is written as:

$$U_5 = U_4 \exp\Big(-ik\Big(\sqrt{d_{\text{po}}^2 + r^2} - d_{\text{po}}\Big)\Big) \times \exp\Big(-ik\Big(\sqrt{d_{\text{sp}}^2 + r^2} - d_{\text{sp}}\Big)\Big) T_{\text{aberr}}(x, y), \tag{5.16}$$

The next step is to apply the RSDI one more to acquire the light field on the image sensor plane, U_6, which is written as

$$U_6(x, y) = \frac{1}{i\lambda} \int_{-\infty}^{\infty} \int_{-\infty}^{\infty} U_5(\xi, \eta) \exp(ikr_{\text{eculi}}) \frac{d_{\text{sp}} + d_{\text{ss}}}{r_{\text{eculi}}^2} \mathrm{d}\xi \mathrm{d}\eta. \tag{5.17}$$

In Eq. (5.17), d_{ss} represents an axial shift of the image sensor from the conjugate point of the IPS. This subtle shift affects the distinguishability between different posture parameters (i.e., different TFM positions and orientations) based on the final signal image. The details of this effect will be discussed later.

The intensity distribution on the image sensor is proportional to the square of the magnitude of U_6, denoted as $I_{\text{pmc}}(x, y; \lambda)$, i.e.

$$I_{\text{pmc}}(x, y; \lambda) \propto |U_6(x, y)|^2. \tag{5.18}$$

In Eq. (5.18), the λ separated by a semicolon indicates the intensity distribution corresponding to a specific wavelength, and the subscript "pmc" stands for "point", "monochromatic", and "coherent".

If the light source is broadband, the intensity distribution on the image sensor is simulated by assuming its spectrum. Suppose the IPS has a uniformly distributed spectrum from λ_{short} to λ_{long}, a series of wavelengths can be expressed as

$$\lambda_m = \lambda_{\text{short}} + m\Delta\lambda. \tag{5.19}$$

Equation (5.19) can be chosen to implement the numerical simulation, where

$$\begin{cases} \Delta\lambda = (\lambda_{\text{long}} - \lambda_{\text{short}})/M, & M \in \mathbb{N} \\ m = 0 \sim M, & m \in \mathbb{N} \end{cases}. \tag{5.20}$$

With Eqs. (5.19) and (5.20), the intensity distribution on the image sensor caused by a broadband and incoherent point source is denoted as I_{pbi}, and it can be written as

$$I_{\text{pbi}}(x, y) = \sum_{m=0}^{M} I_{\text{pmc}}(x, y; \lambda_m), \tag{5.21}$$

where the subscript "pbi" means "point", "broadband", and "incoherent". Finally, if the finite size of the light source is also taken into consideration, the intensity distribution on the image sensor is denoted as $I_{\text{fbi}}(x, y)$, and it can be acquired by a convolution operation expressed by

$$I_{\mathrm{fbi}}(x, y) = I_{\mathrm{pbi}}(x, y) * G_{\mathrm{ps}}(x, y), \tag{5.22}$$

where the asterisk denotes convolution operation and $G_{\mathrm{ps}}(x, y)$ is the geometric projection of the source intensity distribution onto the image sensor plane. The subscript "fbi" means "finite size", "broadband", and "incoherent". The model derived above can generate simulated signal images corresponding to desired posture parameters. The signal images are then called "diffractive images (DIs)".

5.3 DIs with a Horizontal Flat Mirror

In this section, θ_{tilt} and ϕ_{tilt} are set to be zero to keep the TFP horizontal. First, the optical system is assumed to be aberration-free. Next, d_{f} is set from $-30\,\mu\mathrm{m}$ to $+30\,\mu\mathrm{m}$ at an increment of $1\,\mu\mathrm{m}$ to acquire a series of DIs. To simulate realistic DIs, all parameters of the optical system are chosen according to an actual experiment setup, and they are listed in Table 5.1.

Based on the above parameters and the developed model, several DIs acquired by various chosen d_{f} are shown in Fig. 5.3. However, the obvious problem is that the DIs seem to be highly symmetric on both sides of the case of $d_{\mathrm{f}} = 0$.

Table 5.1 Parameter list for DI simulation, in which the flat mirror is fixed horizontally and the optical system is aberration-free

Items	Values	Units
Object side NA	0.28	–
d_{po}	20	mm
d_{sp}	200	mm
d_{ss}	0	mm
R_{p}	5.833	mm
Source width (square shape)	22.8	μm
Wavelengths	0.45, 0.50, 0.55, 0.60, 0.65 (uniform distribution)	μm

Fig. 5.3 Simulated DIs with several chosen d_{f}

The symmetry problem is analyzed quantitatively using the NCC algorithm described by Eq. (4.1) [2]. By calculating the NCC values between two arbitrary simulated DIs, a "similarity matrix" can be obtained, as shown in Fig. 5.4. In this matrix, the color of each point represents the similarity between two DIs corresponding to the abscissa and the ordinate of the point, respectively. Four particular cross-sections, CS1, CS2, CS3, and CS4, of this similarity matrix, are depicted in Fig. 5.5. These curves, called the similarity response curves (SRCs), are similar to the depth response curve (DRC) in conventional confocal microscopy because they describe the relationship between similarity and axial position. Therefore, the peak position of the SRCs can also be used to determine the axial position of a detected surface. However, in Fig. 5.5, the two-peak feature of CS1, CS2, and CS3 indicates the existence of two almost identical DIs, making it impossible to determine which one is correct. This phenomenon vanishes in the case of CS4, which corresponds to the case of $d_f = 0$, but the peak becomes quite "blunt", meaning that peak determination can be less accurate and unstable. This condition is also observable by looking at the "*x*-shape" inside the square region in Fig. 5.4, where a small region shows a similarity distribution resembling a "plateau".

To break the ambiguity caused by the symmetricity mentioned above, some aberration is required to break this symmetricity. Here, the astigmatism is selected as an example to be simulated and discussed. To do this, the aberration term described by Eq. (5.3) is designed to have the form of Zernike polynomial Z_2^2, which is

$$T_{\mathrm{aberr}}(x, y) = c_a r^2 \cos(2\varphi) \tag{5.23}$$

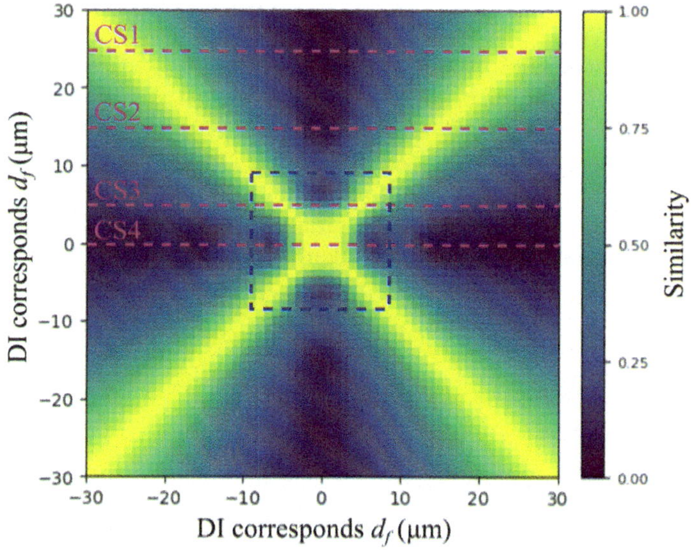

Fig. 5.4 A matrix that reveals the similarities between two arbitrary DIs. Four particular cross-sections CS1, CS2, CS3, and CS4 are depicted as Fig. 5.5

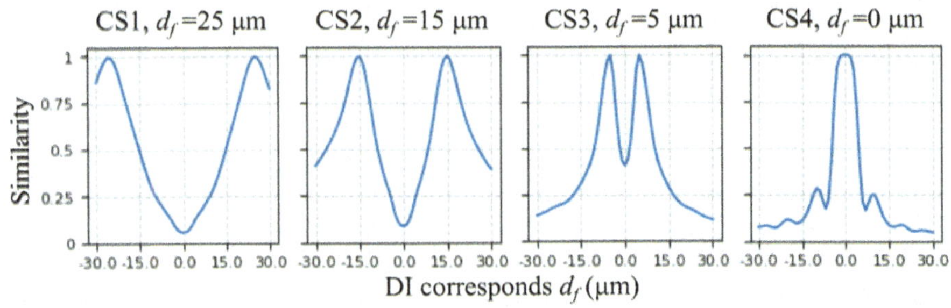

Fig. 5.5 Four cross-sections that correspond to CS1, CS2, CS3, and CS4 in Fig. 5.4

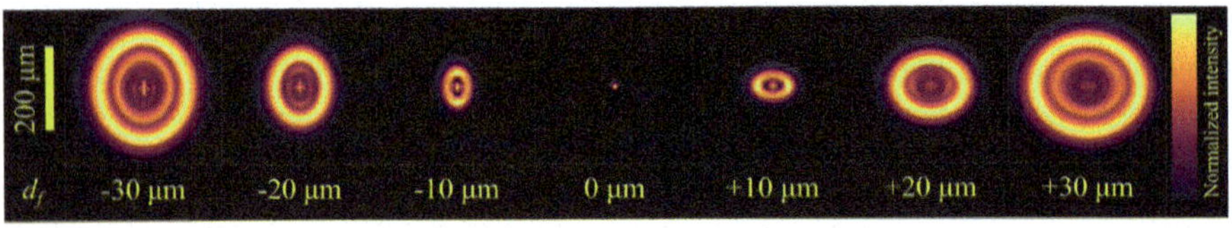

Fig. 5.6 Simulated DIs with several chosen d_{f}

This aberration is commonly called "vertical astigmatism". In which $r = \sqrt{x^2 + y^2}$ and $\phi = \tan^{-1}(y/x)$. c_{a} is a coefficient to adjust the degree of aberration, which is set as $\pi/4R_{\mathrm{p}}^2$ so that the maximum phase deviation on the edge of the pupil function is $\pi/4$. With this extra aberration, DIs with the same d_{f} as chosen in Fig. 5.3 are illustrated in Fig. 5.6. The existing symmetricity can be effectively broken.

This phenomenon is observed in the similarity matrix, as shown in Fig. 5.7, and its cross-sections, as shown in Fig. 5.8. Also, note that the "x-shape" plateau in Fig. 5.4 disappears automatically. It means the extra aberration provides the most basic distinguishability. Therefore, with proper data processing, the position of the horizontal mirror can be inversely inferred.

However, aberration can cause a more prominent focal spot on a detected object. Thus, an extra aberration should be sufficient to break the symmetricity instead of unnecessarily large.

5.4 DIs with a Tilted Flat Mirror

As learned from the previous section, the optical system is designed to have the same astigmatism described by Eq. (5.23) as $c_{\mathrm{a}} = \pi/4R_{\mathrm{p}}^2$ to avoid ambiguity along the optical axis. Cases with chosen posture parameters are simulated based on this precondition, as shown in Fig. 5.9. Unfortunately, a new ambiguity is discovered this time. For example, as shown in Fig. 5.10, the case of $d_{\mathrm{f}} = -20\,\mu\mathrm{m}$, $\theta_{\mathrm{tilt}} = 9°$ and

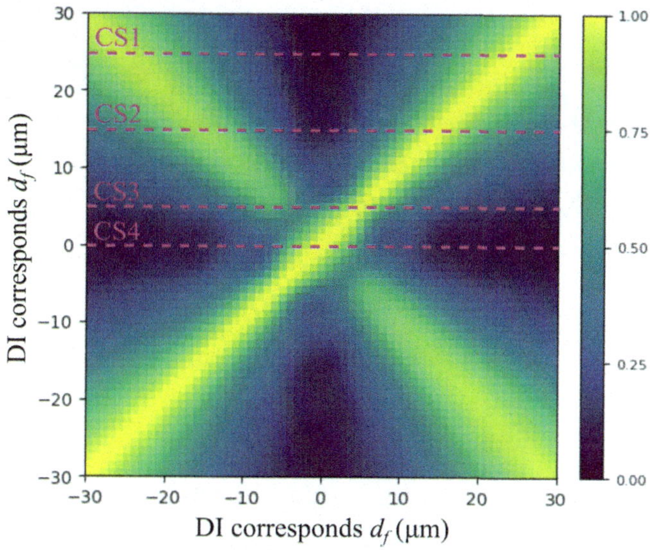

Fig. 5.7 A matrix that reveals the similarities between two arbitrary DIs. Four particular cross-sections CS1, CS2, CS3, and CS4 are depicted as Fig. 5.8

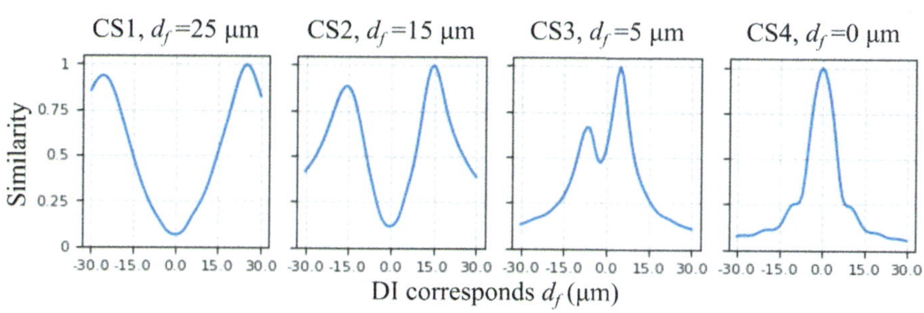

Fig. 5.8 Four cross-sections that correspond to CS1, CS2, CS3, and CS4 in Fig. 5.7

$\phi_{\text{tilt}} = 0°$ is almost the same as the case of $d_{\text{f}} = -20\,\mu\text{m}$, $\theta_{\text{tilt}} = 9°$ and $\phi_{\text{tilt}} = 180°$. Their NCC value is as high as 0.988, which is concluded as indistinguishable.

To break the above ambiguity, the special shifting d_{ss} in Fig. 5.11 is set as $-1000\,\mu\text{m}$ (equivalent to 10 µm defocus at the TFM side). In this case, the same chosen posture parameters are simulated again, as shown in Figs. 5.11 and 5.12. By this time, the ambiguity is broken by an obvious position difference. And their NCC value drops to 0.130. Therefore, the distinguishability of DIs with different posture parameters is guaranteed with extra aberration and special shifting [3].

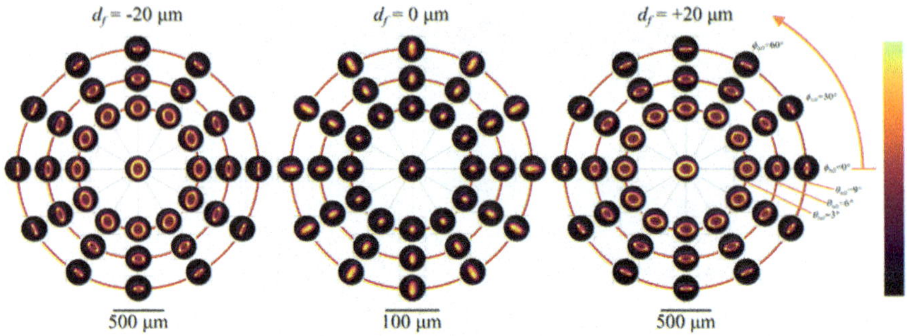

Fig. 5.9 DIs versus various posture parameters with $d_{ss} = 0\,\mu m$

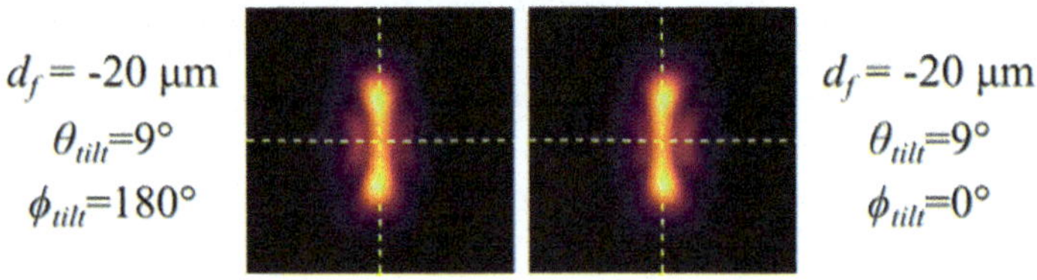

Fig. 5.10 Two DIs that are difficult to distinguish

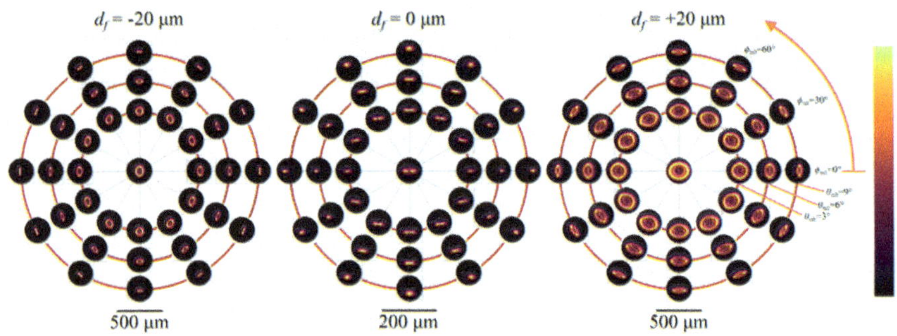

Fig. 5.11 DIs versus various posture parameters with $d_{ss} = -1000\,\mu m$

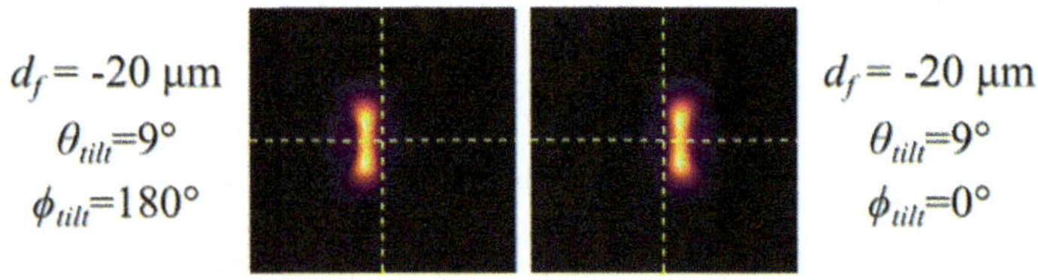

Fig. 5.12 Two DIs that are easier to distinguish due to their different position

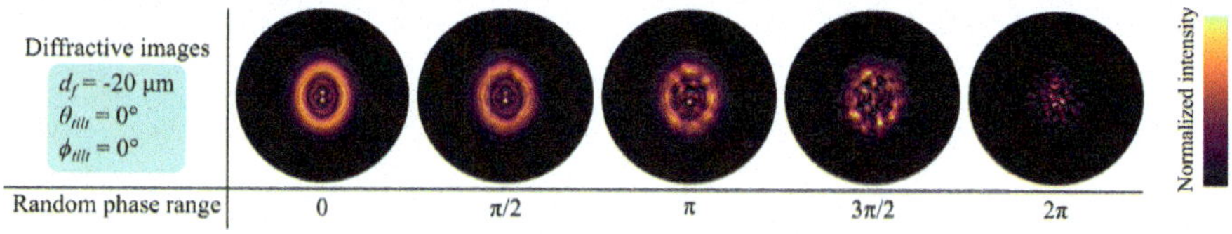

Fig. 5.13 Influence of surface roughness on the DIs

5.5 Nonspecular Flat Sample

To study the influence of a nonspecular flat sample, the transmission function of the TFM expressed as Eq. (5.12) can be multiplied by an additional term that describes the phase deviation caused by surface roughness. Take the case of $d_f = -20\,\mu m$, $\theta_{tilt} = 0°$, $\phi_{tilt} = 0°$, and $d_{ss} = 0$ as an example, the simulated DIs are illustrated in Fig. 5.13. The additional random phase deviations from the left to the right are 0, $\pi/2$, π, $3\pi/2$, 2π respectively. The DIs can be destroyed when the roughness is too high. In this case, inferring the posture parameter according to the DIs may not be achievable. However, practical criteria about the maximum applicable roughness can be determined by designed experiments [4].

References

1. Wu G-W, Chen L-C (2023) Resolving measurement ambiguity in diffractive image microscopy for 6DOF surface measurement using designed aberration and multiple-layer perceptron. Int J Adv Manuf Technol. https://doi.org/10.1007/s00170-023-11147-0
2. Wu G-W, Nguyen DT, Chen L-C (2019) Novel full-field micro surface profilometry using diffractive image correlation without vertical scanning. Opt Lett 44(14):3534–3537
3. Wu G-W, Chen L-C (2021) Precise 3-D microscopic profilometry using diffractive image microscopy and artificial neural network in single-exposure manner. Opt Lasers Eng 147:106732
4. Wu G-W (2022) Theory and implementation of microscopic diffractive image profilometry. Ph.D. thesis, National Taiwan University

Chapter 6
Diffractive Image Profilometry (DIP)

6.1 Introduction

This chapter first introduces the skeleton and the core principle of a diffractive image profilometry (DIP) system. Secondly, the term "database" is defined in detail and explains three possible ways to use a database.

6.2 The Skeleton and Core Principle of a DIP System

A conventional confocal microscope has a pair of conjugate pinholes, an illumination pinhole, and a detection pinhole. The illumination pinhole provides a point illumination, and the detection pinhole serves as a spatial filter, which blocks the light from elsewhere but the detection point. With this approach, well-known optical sectioning is achieved while measuring samples. However, when the detected object has a specular surface, optical sectioning is not necessary, and the detection pinhole blocks some useful information that can tell the surface orientation. Therefore, removing the detection pinhole in a conventional confocal microscope is the first step. The next step is the replacement of the photodetector behind the detection pinhole by an image sensor. This image sensor can record the detailed light distribution, not just a single intensity value. In the detailed light distribution, much information is hidden. After the above modification, a conventional confocal microscope is modified to a system shown in Fig. 6.1, which is called a DIP system. For such a system, the quantities being measured are the position and orientation of the local surface of the underlying object. Furthermore, the raw signal is the DI captured by the image sensor. By analyzing the captured DI, the position and orientation of the local surface can be inferred inversely.

If the distance from the local surface to the ideal focal point, d_f, is small enough, then the illuminated region is also small enough to be approximated by a tangent

L. Chen et al., *Diffractive Image Microscopy for 3D Imaging*, Springer Tracts in Mechanical Engineering, https://doi.org/10.1007/978-981-97-7782-2_6

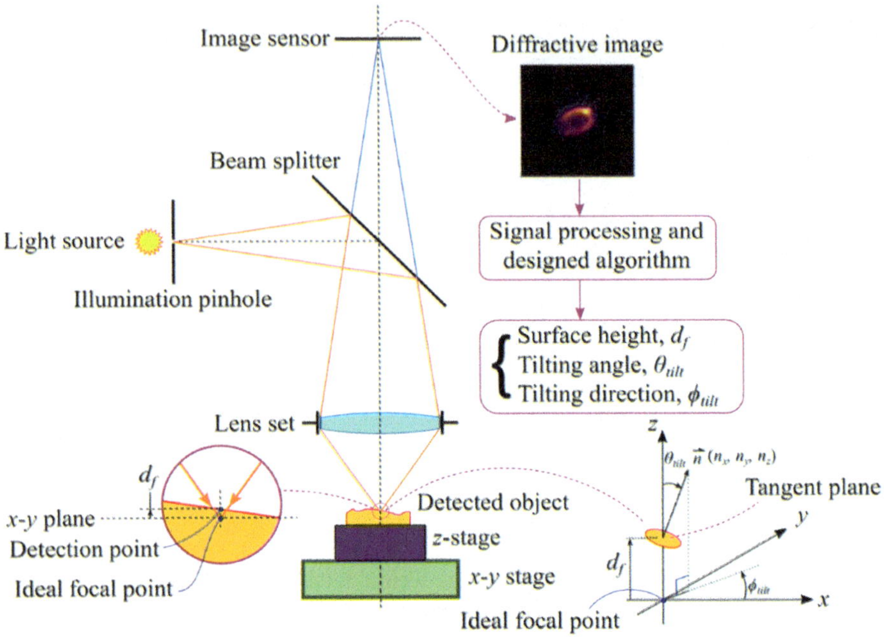

Fig. 6.1 The configuration of a diffractive image profilometer

plane. Therefore, the physical model derived in Chap. 5 can be applied. Similar assumptions can also be found in relevant studies [1, 2]. The three posture parameters can fully define this tangent plane d_f, θ_{tilt}, and ϕ_{tilt}, as mentioned in Sect. 5.2. They correspond to the surface height, tilting angle, and tilting direction, respectively, where the surface height is defined by the z-position of the tangent plane with respect to the ideal focal point. The tilting angle refers to the angle from the z-axis to $\vec{n}$ the tangent plane's normal vector. Tilting direction is the angle from the x-axis to the projection of $\vec{n}$ on the $x - y$ plane. The core of DIP is the methods that convert DIs to posture parameters, which determines the local surface's position and orientation under detection. Finally, the posture parameters of many points compose a complete surface profile.

As introduced in Chap. 5, some theoretical tools are available to generate simulated DIs based on the given posture parameters and relevant conditions. However, there is no simple analytical method to calculate the posture parameter inversely from observed DIs as several barriers hamper this way. Firstly, relatively simple theories often imply many assumptions and simplifications, making them easier to use but less accurate [3]. In contrast, advanced theories can better describe physical phenomena, but inverse formulations are generally unavailable [4]. Thirdly, even if inverse formulations are available, detailed light field information, including amplitude and phase, is nevertheless required. Still, the phase information is lost during the intensity acquisition except for holographic techniques [5]. Finally, any attempt to do an inverse calculation requires detailed information on the used optical system. It means that a real system must be characterized or modeled in advance, this is evidently another challenging work. Therefore, data-driven approaches such as table looking up methods, mathematical model fitting, or machine learning-based methods are relatively effective.

Nevertheless, this does not mean that physical theories are not necessary. On the contrary, they are essential tools to check the existence of one-to-one relationships done in Chap. 5. If the rigorous one-to-one relationship does not exist, then data-driven methods may just bring more confusion and headache. Besides, any data-driven approach requires a high-quality data set as its foundation, discussed in Sect. 7.8. This chapter focuses on the definition of a database and its usage.

6.3 The Definition of a Database

The term "database" in the content means a set of DIs or their extracted features (EFs) labeled by the corresponding posture parameters. However, because of aberration and other imperfections, the optical performance is not uniform across the whole FOV in a real optical system. In other words, the database is actually field-dependent. So first, this means that if the position of the illumination pinhole changes, both the position of the detection point and the manner that DIs or EFs vary with posture parameters change. Second, another factor affecting the DIs or EFs is the geometry of illumination pinholes. If the size or shape of the illumination pinhole changes, the result DIs or EFs change accordingly. Third, DIs are captured by an image sensor and the so-called DIs are a tiny region cropped from the whole image. As illustrated in Fig. 6.2, if the cropping region of a given measurement point changes, the DI can be shifted, trimmed, or even incorrectly selected. Therefore, the cropping region corresponding to a specific field point must be fixed accordingly. The center position of the cropping region of a particular DI is determined by the geometric projection of its source pinhole and the width of the cropping region should be wide enough to include the whole DI.

Conclude from the above; a database can be defined by a set of DIs with corresponding posture parameters and clear specified pinhole or field position, pinhole geometry, and cropping region on the image sensor. For convenience, a database based on DIs is called an image database (IDB), and databases based on EFs have different abbreviations as will be seen later.

In order to clearly explain the followed algorithm, a set of database with a particular field point, pinhole geometry and cropping region is now denoted as $\mathbf{D}_{i,j,k}$, which corresponding to posture parameters $d_{\mathrm{f},i}$, $\theta_{\mathrm{tilt},j}$, and $\phi_{\mathrm{tilt},k}$. For better understanding, Table 6.1 provides a visualized representation of such a database. Besides, the extent of d_{f} also determines the maximum axial measurable range.

Commonly, $d_{\mathrm{f},i}$, $\theta_{\mathrm{tilt},j}$, and $\phi_{\mathrm{tilt},k}$ are sampled in an equal space manner, with sampling pitch P_{df}, P_{ta}, and P_{td}, respectively. This means that

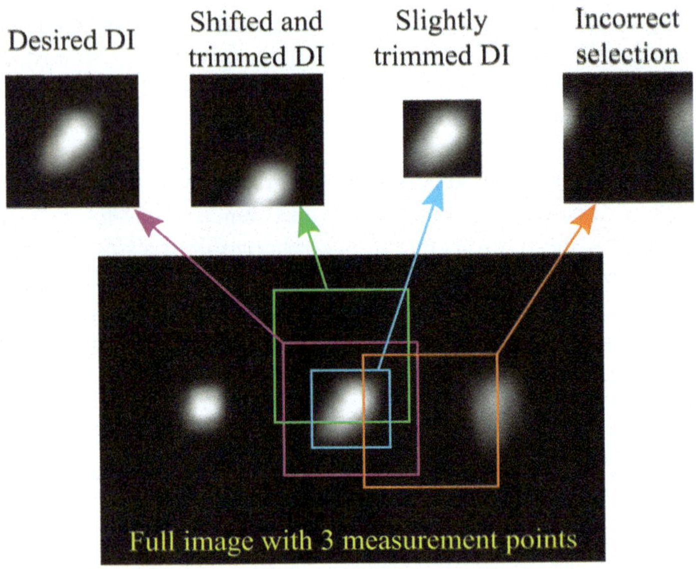

Fig. 6.2 The cropping region must be fixed for a given measurement point. Otherwise, the cropped DI will change accordingly

Table 6.1 A set of database presented as a table

Diffractive images DIs ($\mathbf{D}_{i,j,k}$)	Corresponding posture parameters		
	Surface height $d_{f,i}$	Tilting angle, $\theta_{tilt,j}$	Tilting direction $\phi_{tilt,k}$
$\mathbf{D}_{1,1,1}$	$d_{f,1}$	$\theta_{tilt,1}$	$\phi_{tilt,1}$
$\mathbf{D}_{1,1,2}$	$d_{f,1}$	$\theta_{tilt,1}$	$\phi_{tilt,2}$
$\mathbf{D}_{1,1,3}$	$d_{f,1}$	$\theta_{tilt,1}$	$\phi_{tilt,3}$
…	…	…	…
$\mathbf{D}_{1,2,1}$	$d_{f,1}$	$\theta_{tilt,2}$	$\phi_{tilt,1}$
$\mathbf{D}_{1,2,2}$	$d_{f,1}$	$\theta_{tilt,2}$	$\phi_{tilt,2}$
$\mathbf{D}_{1,2,3}$	$d_{f,1}$	$\theta_{tilt,2}$	$\phi_{tilt,3}$
…	…	…	…
$\mathbf{D}_{2,1,1}$	$d_{f,2}$	$\theta_{tilt,1}$	$\phi_{tilt,1}$
$\mathbf{D}_{2,1,2}$	$d_{f,2}$	$\theta_{tilt,1}$	$\phi_{tilt,2}$
$\mathbf{D}_{2,1,3}$	$d_{f,2}$	$\theta_{tilt,1}$	$\phi_{tilt,3}$
…	…	…	…

$$\begin{cases} d_{f,i+1} = d_{f,i} + P_{df} \\ \theta_{tilt,j+1} = \theta_{tilt,j} + P_{ta} \\ \phi_{tilt,k+1} = \phi_{tilt,k} + P_{td} \end{cases} . \tag{6.1}$$

With this sampling principle, the total number of DIs within one database can be calculated according to

$$N_{\mathrm{DI}} = \left\lfloor \frac{R_{\mathrm{df}}}{P_{\mathrm{df}}} + 1 \right\rfloor \left\lfloor \frac{R_{\mathrm{ta}}}{P_{\mathrm{ta}}} \right\rfloor \left\lfloor \frac{R_{\mathrm{td}}}{P_{\mathrm{td}}} \right\rfloor + \left\lfloor \frac{R_{\mathrm{df}}}{P_{\mathrm{df}}} + 1 \right\rfloor, \tag{6.2}$$

where N_{DI} is the total number of DIs within the database. R_{df}, R_{ta}, and R_{td} are the sampling range for d_{f}, θ_{tilt}, and ϕ_{tilt}, respectively. The notation $\lfloor\rfloor = 1$ is the floor function, which takes the largest integer that does not exceed the value inside. For example, $\lfloor 1.23 \rfloor = 1$ and $\lfloor 5.0 \rfloor = 5$. In Eq. (6.2), the inconsistent expression over the three parameters is due to the following two facts:

(1) When $\theta_{\mathrm{tilt}} = 0$, the local surface is just a horizontal plane, which means any ϕ_{tilt} leads to the same DI.
(2) The tilting direction, ϕ_{tilt}, is a cyclic parameter, i.e., the case of $\phi_{\mathrm{tilt}} = 0°$ is as same as the case of $\phi_{\mathrm{tilt}} = 360°$.

6.4 Reduced Database

A complete database is rather bulky, and it takes much effort if databases corresponding to multiple field points are desired. However, if the object to be measured only contains 2.5-D features, i.e., all local surfaces are horizontal, just like a step height standard. In this case, a reduced database can be built by restricting θ_{tilt} and ϕ_{tilt} zero, only leaving d_{f} as a variable. Such a reduced database is denoted as $\mathbf{D}_i$ and which is only labeled by $d_{\mathrm{f},i}$. In this case, Eq. (6.2) reduces to

$$N_{\mathrm{DI}} = \left\lfloor \frac{R_{\mathrm{df}}}{P_{\mathrm{df}}} + 1 \right\rfloor. \tag{6.3}$$

Therefore, the total number of DIs drops dramatically, and the database building process can become much more manageable.

6.5 Possible Ways of Using a Database

6.5.1 Image Comparison Method

Suppose a set of databases is already being built and denoted as the description in Sect. 6.3. Then, the most direct way to use the database in real measurement is to look for the most similar DI in this database. This can be done by NCC algorithm introduced in Sect. 4.2. Now let $\mathbf{M}$ be an observed DI in a real measurement, then the similarities between $\mathbf{M}$ and $\mathbf{D}_{i,j,k}$ are calculated and denoted as $s_{i,j,k}$. The most elementary usage is looking for a set of indices $\hat{\imath}, \hat{\jmath}, \hat{k}$ that $s_{\hat{\imath},\hat{\jmath},\hat{k}}$ has the maximum value. Then the posture parameters of the local surface under observation are inferred to be $d_{\mathrm{f},\hat{\imath}}$, $\theta_{\mathrm{tilt},\hat{\jmath}}$, and $\phi_{\mathrm{tilt},\hat{k}}$. This is called the maximum NCC-looking method (MNCC). If the items to be compared are not DIs but their EFs, then the looking target is a point with the highest ZNCC value; in this case, the method is called the maximum ZNCC-looking method (MZNCC).

Due to discrete sampling, posture parameters inferred by the above method can only be a particular sampling point within the database. However, the posture parameters are generally in between several sampling points for a real local surface. This means that the sampling pitch of the database limits the resolution. In order to improve the resolution, either a denser database or a specific interpolation is required. This is a three-dimensional interpolation problem in $d_\mathrm{f} - \theta_\mathrm{tilt} - \phi_\mathrm{tilt}$ space with a single output, the similarity, s.

To deal with this problem, the point with the highest similarity, $(d_{\mathrm{f},\hat{\imath}}, \theta_{\mathrm{tilt},\hat{\jmath}}, \phi_{\mathrm{tilt},\hat{k}})$, must be determined first. Second, adjacent sampling points within a volume defined by

$$\begin{cases} \left|d_\mathrm{f} - d_{\mathrm{f},\hat{\imath}}\right| \le H_\mathrm{df} \\ \left|\theta_\mathrm{tilt} - \theta_{\mathrm{tilt},\hat{\jmath}}\right| \le H_\mathrm{ta} \\ \left|\phi_\mathrm{tilt} - \phi_{\mathrm{tilt},\hat{k}}\right| \le H_\mathrm{td} \end{cases}, \tag{6.4}$$

are selected, where H_df, H_ta, and H_{td} are predetermined half ranges for d_f, θ_tilt, and ϕ_tilt, respectively. For simplicity, the number of sampling points along d_f, θ_tilt, and ϕ_tilt directions are set as the same, denoted as N_s. Where N_s is an odd number and the point with the highest similarity is located at the center. Figure 6.3 shows 27 selected points within the volume defined by Eq. (6.4) when $N_\mathrm{s} = 3$, in which P_df, P_ta, and P_td are the sampling pitch of d_f, θ_tilt, and ϕ_tilt, respectively. Although Fig. 6.3 depicts the case, the minimum value is typically 5 to ensure enough selected points so that the following fitting procedure can be surely determined.

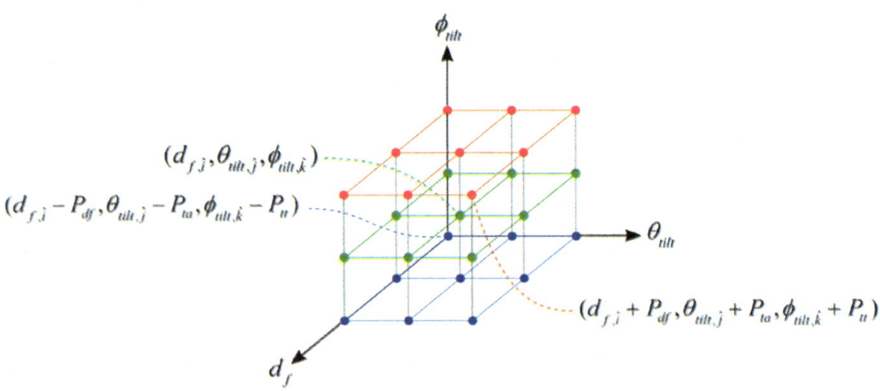

Fig. 6.3 27 selected point for the followed fitting procedure in case of $N_\mathrm{s} = 3$

Next, a $\mathbb{R}^3 \rightarrow \mathbb{R}^1$ function in the form of $s = f_{\mathrm{fit}}(d_{\mathrm{f}}, \theta_{\mathrm{tilt}}, \phi_{\mathrm{tilt}})$ needs to be designed to fit the above selected points. A suitable equation for such a high-dimensional problem is not that obvious. However, a fractional polynomial expressed by

$$f_{\mathrm{fit}} = f_{\mathrm{nu}}/f_{\mathrm{de}} \tag{6.5}$$

is tested to be valid for this problem. Where f_{nu} and f_{de} are third-degree polynomials with three variables, they are explicitly written as

$$\begin{aligned} f_{\mathrm{nu}} = & \; b_0 + (b_1 d_{\mathrm{f}} + b_2 \theta_{\mathrm{tilt}} + b_3 \phi_{\mathrm{tilt}}) \\ & + \left[\left(b_4 d_{\mathrm{f}}^2 + b_5 \theta_{\mathrm{tilt}}^2 + b_6 \phi_{\mathrm{tilt}}^2 \right) + \left(b_7 d_{\mathrm{f}} \theta_{\mathrm{tilt}} + b_8 \theta_{\mathrm{tilt}} \phi_{\mathrm{tilt}} + b_9 \phi_{\mathrm{tilt}} d_{\mathrm{f}} \right) \right] \\ & + \left[\left(b_{10} d_{\mathrm{f}}^3 + b_{11} \theta_{\mathrm{tilt}}^3 + b_{12} \phi_{\mathrm{tilt}}^3 \right) \right. \\ & + \left(b_{13} d_{\mathrm{f}}^2 \theta_{\mathrm{tilt}} + b_{14} d_{\mathrm{f}} \theta_{\mathrm{tilt}}^2 + b_{15} \theta_{\mathrm{tilt}}^2 \phi_{\mathrm{tilt}} + b_{16} \theta_{\mathrm{tilt}} \phi_{\mathrm{tilt}}^2 + b_{17} \phi_{\mathrm{tilt}}^2 d_{\mathrm{f}} + b_{18} \phi_{\mathrm{tilt}} d_0^2 \right) \\ & \left. + b_{19} d_{\mathrm{f}} \theta_{\mathrm{tilt}} \phi_{\mathrm{tilt}} \right], \end{aligned} \tag{6.6}$$

and

$$\begin{aligned} f_{\mathrm{de}} = & \; c_0 + (c_1 d_{\mathrm{f}} + c_2 \theta_{\mathrm{tilt}} + c_3 \phi_{\mathrm{tilt}}) \\ & + \left[\left(c_4 d_{\mathrm{f}}^2 + c_5 \theta_{\mathrm{tilt}}^2 + c_6 \phi_{\mathrm{tilt}}^2 \right) + \left(c_7 d_{\mathrm{f}} \theta_{\mathrm{tilt}} + c_8 \theta_{\mathrm{tilt}} \phi_{\mathrm{tilt}} + c_9 \phi_{\mathrm{tilt}} d_{\mathrm{f}} \right) \right] \\ & + \left[\left(c_{10} d_{\mathrm{f}}^3 + c_{11} \theta_{\mathrm{tilt}}^3 + c_{12} \phi_{\mathrm{tilt}}^3 \right) \right. \\ & + \left(c_{13} d_{\mathrm{f}}^2 \theta_{\mathrm{tilt}} + c_{14} d_{\mathrm{f}} \theta_{\mathrm{tilt}}^2 + c_{15} \theta_{\mathrm{tilt}}^2 \phi_{\mathrm{tilt}} + c_{16} \theta_{\mathrm{tilt}} \phi_{\mathrm{tilt}}^2 + c_{17} \phi_{\mathrm{tilt}}^2 d_{\mathrm{f}} + c_{18} \phi_{\mathrm{tilt}} d_{\mathrm{f}}^2 \right) \\ & \left. + c_{19} d_{\mathrm{f}} \theta_{\mathrm{tilt}} \phi_{\mathrm{tilt}} \right]. \end{aligned} \tag{6.7}$$

There are 40 parameters ($b_0 \sim b_{19}$ and $c_0 \sim c_{19}$) in Eqs. (6.6) and (6.7), and the suggested initial parameters are:

(1) $c_0 = 1$.
(2) b_0 is set as the maximum NCC (or ZNCC) value.
(3) Other parameters are all zeros.

In the case of $N_{\mathrm{s}} = 5$, this fitting problem is generally overdetermined because the number of parameters is 40, but the number of data points is $N_{\mathrm{s}}^3 = 125$. Although this fitting problem seems terrible, it can be done within 0.2 s without special computation caution on a normal office computer (Intel i7-9700 CPU, single thread computation). Let the final fitted function be $s = \hat{f}_{\mathrm{fit}}(d_{\mathrm{f}}, \theta_{\mathrm{tilt}}, \phi_{\mathrm{tilt}})$, then the measured posture parameters are given by

$$\hat{d}_{\mathrm{f}}, \hat{\theta}_{\mathrm{tilt}}, \hat{\phi}_{\mathrm{tilt}} = \underset{d_{\mathrm{f}}, \theta_{\mathrm{tilt}}, \phi_{\mathrm{tilt}}}{\arg\max} \hat{f}_{\mathrm{fit}}(d_{\mathrm{f}}, \theta_{\mathrm{tilt}}, \phi_{\mathrm{tilt}}) \tag{6.8}$$

within the volume that contains the selected points. The result acquired by this fitting-based method is interpolated, thus it is called the interpolated NCC method (INCC). Similarly, suppose the items to be compared are EFs instead of DIs. In that case, the method is called the interpolated ZNCC method (IZNCC) because the ZNCC algorithm is used for computing the similarities between EFs.

As an easy-understanding summary, Fig. 6.4 shows a fitting example done by the LMA. First, the procedure described in Chap. 4 is used to generate simulated DIs. Then one particular DIs is selected as the observed DI and others as the database DIs. Next, the similarity between the observed DI and all database DIs are calculated to get $s_{i,j,k}$, which is then fitted by Eq. (6.5). The fitting result is shown in Fig. 6.4, in which Fig. 6.4a–c represent three hyper cross-sections of the fitted function, $\hat{f}_{\mathrm{fit}}(d_{\mathrm{f}}, \theta_{\mathrm{tilt}}, \phi_{\mathrm{tilt}})$. Where these three hyper cross-sections all pass through the point with the highest similarity, $(d_{\mathrm{f},\hat{i}}, \theta_{\mathrm{tilt},\hat{j}}, \phi_{\mathrm{tilt},\hat{k}})$. Be aware that the three curves in Fig. 6.4 are not fitted separately but together in the $d_{\mathrm{f}} - \theta_{\mathrm{tilt}} - \phi_{\mathrm{tilt}}$ space. Limited by dimension, all points involved in the fitting procedure cannot be fully illustrated on the paper.

Relatively simple, if a reduced database is used, the observed DI only needs to compare with the database DI with horizontal orientation which only gives a similarity series s_i instead of $s_{i,j,k}$. In this case, a one-dimensional cubic polynomial written as

$$s = a_0 + a_1 d_{\mathrm{f}} + a_2 d_{\mathrm{f}}^2 + a_3 d_{\mathrm{f}}^3 \tag{6.9}$$

is sufficient to fit s_i and find the maximum points, with only 4 coefficients $a_0 \sim a_3$ to be determined. As a summary of this section, the above description condenses to a flowchart shown in Fig. 6.5.

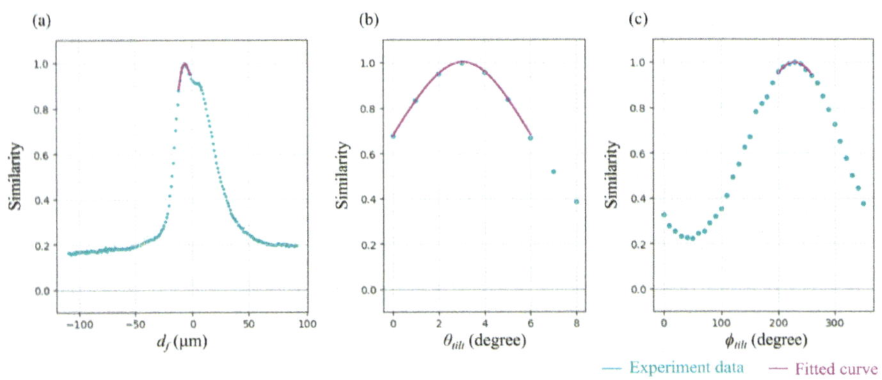

Fig. 6.4 The fitting result to a set of generated data. **a** A cross-section that is parallel to the $s - d_{\mathrm{f}}$ hyperplane. **b** A cross-section that is parallel to the $s - \theta_{\mathrm{tilt}}$ hyperplane. **c** A cross-section that is parallel to the $s - \phi_{\mathrm{tilt}}$ hyperplane

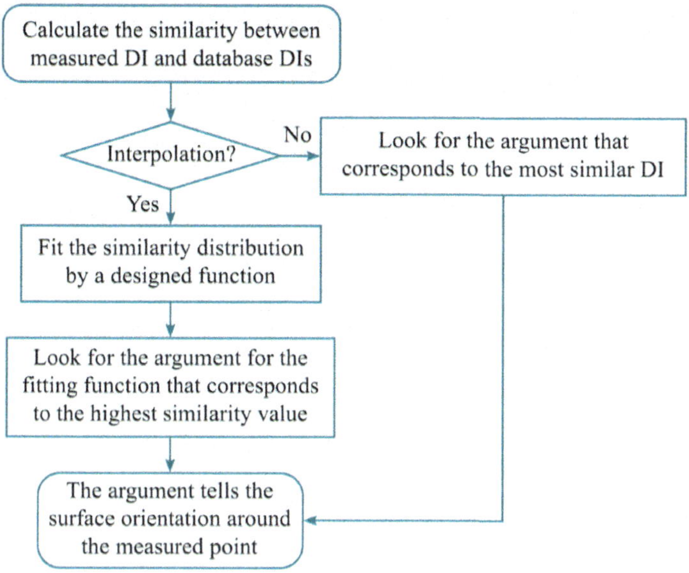

Fig. 6.5 Flowchart of image comparison method

6.5.2 Feature Comparison Method

In the previous section, the original DIs is used to calculate the similarity data, $s_{i,j,k}$. However, it is possible to describe a DI by several extracted features instead of the original image, which typically contains hundreds up to thousands of pixels. Such a process can be seen as a kind of dimensionality reduction, and there are several advantages to doing this. First, the size of the database can be dramatically decreased. Second, the computation loading during real measurement is reduced. Third, it is also a beneficial preprocessing for machine learning-based approaches that will be introduced in the next section.

Before extracting the features, the original DIs must be normalized according to their maximum Grayscale value (GV) to minimize the influence of light source fluctuation and exposure time. For the normalized DIs, two kinds of methods are proposed. One is to fit the DIs by a designed equation, and the fitted parameters are treated as its features. The other one uses the principal component analysis (PCA), by which a DI can be expressed by two principal components (two-unit eigenvectors and two corresponding eigenvalues) and its centroid position.

Now revisit Figs. 5.9 and 5.11, it can be observed that the DIs of an astigmatic system typically have ellipse contours. Based on this observation, two analytical equations are designed to fit the DIs by the LMA. The first one is a two-dimensional Gaussian surface, which is expressed by

$$\begin{cases} g(m_{\text{di}}, n_{\text{di}}) = A_{\text{g}} \exp\left(-\frac{u^2}{2\sigma_u^2} - \frac{v^2}{2\sigma_v^2}\right) + B_k \\ \begin{bmatrix} u \\ v \end{bmatrix} = \begin{bmatrix} \cos\alpha & \sin\alpha \\ -\sin\alpha & \cos\alpha \end{bmatrix} \begin{bmatrix} m_{\text{di}} - \mu_m \\ n_{\text{di}} - \mu_n \end{bmatrix} \end{cases}. \tag{6.10}$$

And the second one is a squared fractional polynomial that expressed by

$$\begin{cases} g(m_{\text{di}}, n_{\text{di}}) = \left[\frac{g_0}{1+g_1u^2+g_2v^2}\right]^2 + B_k \\ \begin{bmatrix} u \\ v \end{bmatrix} = \begin{bmatrix} \cos\alpha & \sin\alpha \\ -\sin\alpha & \cos\alpha \end{bmatrix} \begin{bmatrix} m_{\text{di}} - \mu_m \\ n_{\text{di}} - \mu_n \end{bmatrix} \end{cases}, \tag{6.11}$$

Refer to the coordinate definition in Fig. 6.6a, common coefficients in Eqs. (6.10) and (6.11) are introduced first. In which m_{di} and n_{di} are the row and column coordinates respectively. $g(m_{\text{di}}, n_{\text{di}})$ is the GV at point $(m_{\text{di}}, n_{\text{di}})$. (μ_m, μ_n) is the coordinate of the center of the fitting function. u and v are two orthogonal but rotated axes that parallel the "long-axis" and the "short-axis" respectively, both axes pass through (μ_m, μ_n). α is the angle from the m_{di}-axis to the u-axis where B_k is a GV offset coefficient. Specifically for Eq. (6.10), A_{g} is the amplitude of the Gaussian surface, σ_u and σ_v represent the standard deviations along the u-axis and v-axis, respectively. Specifically for Eq. (6.11), $g_0 \sim g_2$ are three coefficients to be determined, where g_1 and g_2 are related with the width along the "long-axis" and the "short-axis." All the above parameters and their geometric or image interpretations are listed in Table 6.2.

In order to demonstrate the above fitting methods, a generated DI shown in Fig. 6.6a is chosen as an example to be fitted. Where the fitting result of Eqs. (6.10) and (6.11) are shown in Fig. 6.6b, c, respectively. Both of them provide a satisfactory representation of the original DI. Although there must be some difference between the original DI and the fitting result, anyway, the purpose is to extract the key features, not a perfect fitting.

Very importantly, after the above fitting process, some special treatments to α, σ_u, σ_v, g_1, and g_2 are required. First, due to the square operation to σ_u and σ_v, their

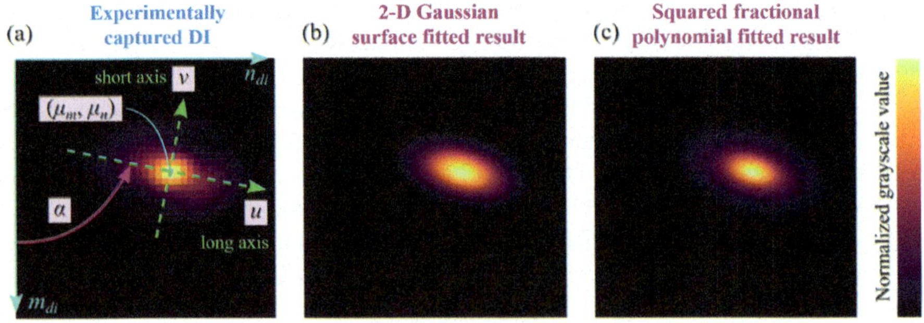

Fig. 6.6 **a** Original DI and coordinate definition. **b** Fitting result of the two-dimensional Gaussian surface. **c** Fitting result of the squared fractional polynomial

Table 6.2 Extracted DI features of fitting-based methods and the corresponding interpretations

DI features	Corresponding feature parameters	
	Two-dimensional Gaussian surface fitting	Squared fractional polynomial fitting
Center position (m_{di}-direction)	μ_m	μ_m
Center position (n_{di}-direction)	μ_n	μ_n
Long-axis direction	α	α
Width along the long-axis	σ_u	g_1(smaller means wider)
Width along the short-axis	σ_v	g_2(smaller means wider)
Amplitude or amplitude related	A_g	g_0
Background (offset)	B_k	B_k

signs must be forced to be positive to ensure a one-to-one relationship. Second, the fitting process cannot distinguish between "long-axis" and "short-axis". This means that σ_u and σ_v may be swapped, g_1 and g_2 also have the same problem. So we need to swap σ_u and σ_v if $|\sigma_u| < |\sigma_v|$. Similarly but reversely, swap g_1 and g_2 if $g_1 > g_2$ because these two parameters are negatively correlated with the corresponding width, as emphasized in Table 6.2.

Next, if σ_u and σ_v are swapped, α should be added with $\pi/2$ then wrap into the range $[0, \pi)$, otherwise only the wrapping is required. Similarly, if g_1 and g_2 are swapped, α should be added with $\pi/2$ then wrap into the range $[0, \pi)$, otherwise only the wrapping is required. The above correction process with regard to Eq. (6.10) is written as

$$\begin{cases} \sigma_u \leftarrow |\sigma_u| \\ \sigma_v \leftarrow |\sigma_v| \end{cases}, \tag{6.12}$$

$$\text{SWAP}(\sigma_u, \sigma_v), \quad \text{if } \sigma_u < \sigma_v, \tag{6.13}$$

$$\begin{cases} \alpha \leftarrow \alpha - \lfloor \alpha/\pi \rfloor \pi, \text{ if no swap} \\ \alpha \leftarrow (\alpha + \pi/2) - \lfloor (\alpha + \pi/2)/\pi \rfloor \pi, \text{ if } \sigma_u \text{ and } \sigma_v \text{ are swapped.} \end{cases} \tag{6.14}$$

The correction process with regard to Eq. (6.11) is written as

$$\text{SWAP}(g_1, g_2), \quad \text{if } g_1 > g_2, \tag{6.15}$$

$$\begin{cases} \alpha \leftarrow \alpha - \lfloor \alpha/\pi \rfloor \pi, \text{ if no swap} \\ \alpha \leftarrow (\alpha + \pi/2) - \lfloor (\alpha + \pi/2)/\pi \rfloor \pi, \text{ if } g_1 \text{ and } g_2 \text{ are swapped.} \end{cases} \tag{6.16}$$

After the above correction, a DI has been "reduced" to its fitting parameters, they are now specially called the feature parameters of this DI. And the feature parameters of a particular DI can be expressed as a vector and denoted as $\mathbf{F}_{i,j,k}$, which corresponding to posture parameters $d_{f,i}$, $\theta_{tilt,j}$, and $\phi_{tilt,k}$. Take the case of Eq. (6.10) as an example. $\mathbf{F}_{i,j,k}$ is explicitly written as

$$\mathbf{F}_{i,j,k} = \left[\mu_{m;i,j,k}, \mu_{n;i,j,k}, \alpha_{i,j,k}, \sigma_{u;i,j,k}, \sigma_{v;i,j,k}, A_{g;i,j,k}, B_{k;i,j,k}\right]^T. \quad (6.17)$$

And very importantly, the suggest initial parameters for the above two fittings are listed in Table 6.3.

Besides the above fitting-based method, PCA can also provide important features of a DI will ellipse contour. However, because an image is not a data type of "a spread of many points", a different point of view is required in order to apply the conventional PCA. Notice that a digital image is recorded as many GVs and those GVs are all integers. Therefore, a GV of 123 at (m_{di}, n_{di}) pixel can be interpreted as a stack of 123 points at the position of (m_{di}, n_{di}) pixel. With this point of view, an image can be converted into a long list of many points. For example, Fig. 6.7 shows a tiny image with 2×2 pixels and it is converted into a list of 7 points, where 7 is just the summation of all GVs. With this interpretation, the covariance matrix of the point list can be calculated. Then, the two eigenvectors and two eigenvalues of the covariance matrix and the centroid of these points can be treated as the features of this tiny image. For example, the same generated DI, as shown in Fig. 6.6a, is interpreted in this way and analyzed by the PCA; the two principal components are drawn in Fig. 6.8 with their relative significance.

Therefore, five independent parameters can be acquired with PCA, they are packed as a feature vector as before, like

$$\mathbf{F}_{i,j,k} = \left[\mu_{m;i,j,k}, \mu_{n;i,j,k}, \alpha_{i,j,k}, e_{1;i,j,k}, e_{2;i,j,k}\right]^T. \quad (6.18)$$

Table 6.3 Suggest initial parameters for DI fitting

Two-dimensional Gaussian surface fitting	Initial value	Squared fractional polynomial fitting	Initial value
μ_m	Half of DI's width	μ_m	Half of DI's width
μ_n	Half of DI's width	μ_n	Half of DI's width
α	0	α	0
σ_u	Tenth of the DI's width	g_1	0.05
σ_v	Tenth of the DI's width	g_2	0.05
A_g	The maximum GV in DI	g_0	Square root of the maximum GV in DI
B_k	0	B_k	0

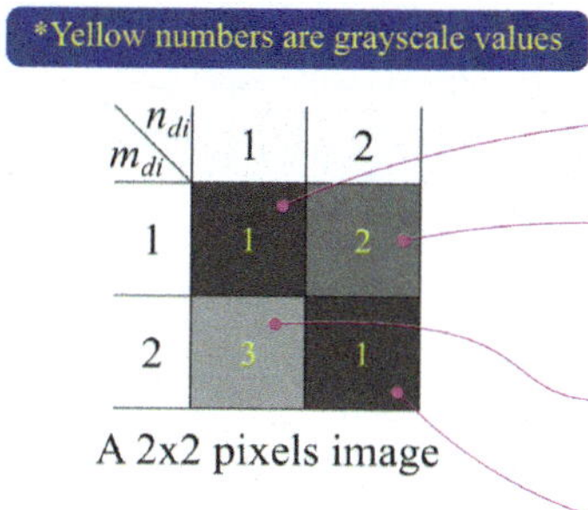

Point No.	Location (m_{di}, n_{di})
1	(1, 1)
2	(1, 2)
3	(1, 2)
4	(2, 1)
5	(2, 1)
6	(2, 1)
7	(2, 2)

Fig. 6.7 A "point list" interpretation of an image

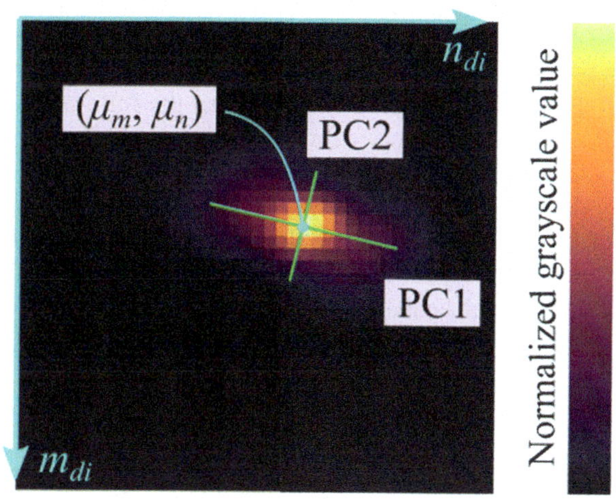

Fig. 6.8 Feature extraction by the principal component analysis. PC1 and PC2 are the first and second principal components, respectively

In Eq. (6.18), $(\mu_{m;i,j,k}, \mu_{n;i,j,k})$ is the centroid position as before. $\alpha_{i,j,k}$ is the direction angle of the first principal component, which is also limited in the range of $[0, \pi)$. $e_{1;i,j,k}$ and $e_{2;i,j,k}$ are the eigenvalues corresponding to the first and the second principal components, respectively. The geometric or image interpretations of these PCA products are listed in Table 6.4.

It is to be noted that the extracted feature vector above is not going to be used directly; the feature normalization introduced in Sect. 4.2 needs to be applied first. Take the first element in Eq. (6.17) as an example. It is normalized according to

$$\mu'_{m;i,j,k} = \frac{\mu_{m;i,j,k} - \mathrm{EXP}(\mu_m)}{\mathrm{STD}(\mu_m)}. \tag{6.19}$$

After normalizing all elements inside Eq. (6.17), it becomes a normalized feature vector (NFV), which is explicitly written as

Table 6.4 Extracted DI features of the PCA method and the corresponding interpretations, in which PC1 and PC2 refer to the first and the second principal components, respectively

DI features	Corresponding feature parameters
	PCA products
Center position (m_{di}-direction)	μ_m
Center position (n_{di}-direction)	μ_n
Long-axis direction	α
Width along the long-axis	e_1
Width along the short-axis	e_1

$$\mathbf{F}'_{i,j,k} = \left[\mu'_{m;i,j,k}, \mu'_{n;i,j,k}, \alpha'_{i,j,k}, \sigma'_{u;i,j,k}, \sigma'_{v;i,j,k}, A'_{g;i,j,k}, B'_{k;i,j,k}\right]^T. \tag{6.20}$$

Now let **M** be an observed DI in a real measurement and its NFV is ω. Then the similarities between ω and the NFV of a database DIs, $\mathbf{F}'_{i,j,k}$, are calculated by ZNCC algorithm and denoted as $s'_{i,j,k}$. Then the followed process to acquire the measured posture parameters is to find the posture parameter with the highest similarity, which is the same as described in Sect. 6.5.1.

Finally, for convenience, a database based on two-dimensional Gaussian surface fitting is called a Gaussian database (GDB). Similarly, a database based on squared fractional polynomial fitting is called a squared fractional polynomial database (FDB), and a database based on PCA is called a PCA database (PDB). As a summary of this section, the above description condenses to a flowchart shown in Fig. 6.9.

6.5.3 Machine Learning-Based Method

As mentioned before, the original DIs or their NFVs are used the calculate the similarities, and if interpolation is desired, equations like Eqs. (6.5)–(6.7) need to be designed. However, there exist situations in manual design that are not that possible. In this case, the machine learning-based approaches provide a different way to solve this high-dimensional fitting problem. When the two-dimensional Gaussian surface fitting is adopted, the NFV is written as Eq. (6.17), and it can be treated as the input vector of an MLP, and the desired outputs of such an MLP are the three posture parameters. However, the hidden layers between the input and output layers are typically determined experimentally. Nevertheless, the number of hidden layers should not exceed 3 to avoid gradient vanishing or exploding [6]. Also, the activation functions of the last hidden layer should be at least a C^1 function to ensure a smooth mapping relationship. Meanwhile, without nitpicking concerns, all activation functions can be selected as the logistic function defined in Eq. (4.19) or the exponential linear unit (ELU) expressed as

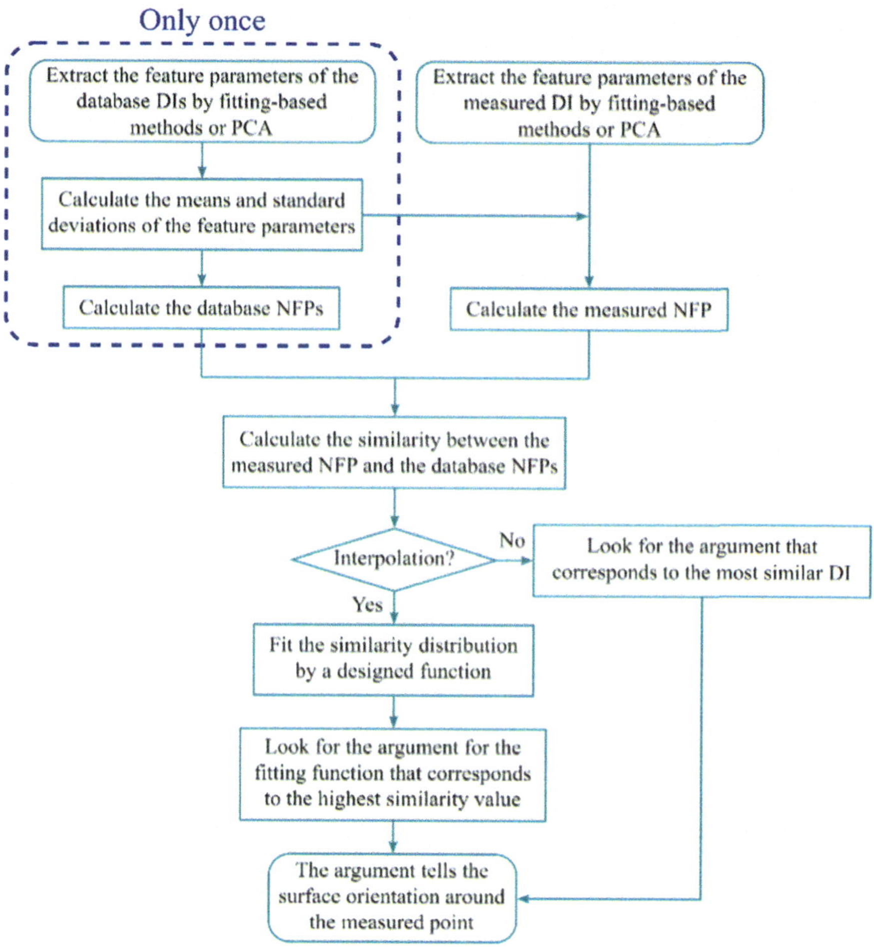

Fig. 6.9 Flowchart of feature comparison method

$$\mathrm{ELU}(\eta) = \begin{cases} \eta, & \eta \geq 0 \\ e^{\eta} - 1, & \eta < 0 \end{cases} \tag{6.21}$$

As mentioned above, a suitable MLP architecture is depicted in Fig. 6.10. In the figure, the dimension of the used NFV and the posture parameters in this dissertation determine the numbers of input and output layer neurons. Furthermore, the number of hidden layers and their neurons must be determined experimentally.

Once the MLP architecture is determined, all the weights and biases are initialized using the Xavier initialization method [7]. Then, a set of prepared NFVs and their corresponding posture parameters are used to train the MLP. It is also suggested to feature normalized posture parameters for better converging, called normalized

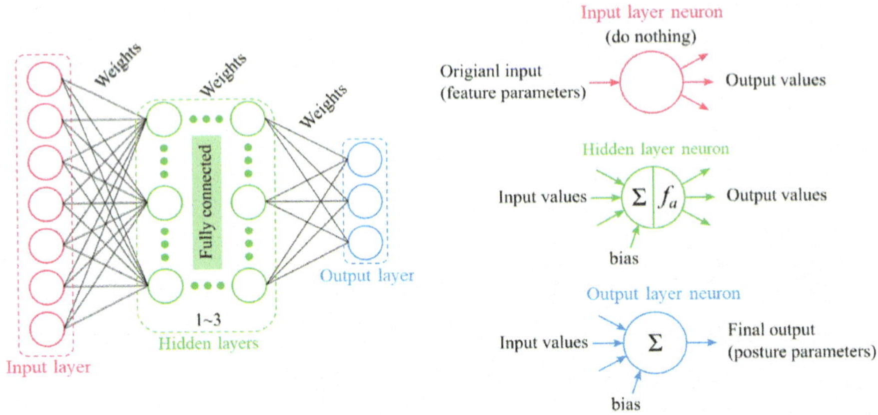

Fig. 6.10 An MLP architecture that can be used to map the normalized feature parameters to the posture parameters

posture parameters (NPP). The next step is the training process, where the following steps are repeated to optimize weights and biases inside the MLP:

(1) Feed the NFVs into the MLP via the input layer.
(2) Forward propagates the NFVs to the output layer to get the final output values.
(3) Compute the SSE between the final output values and the NFVs' corresponding NPP.
(4) Compute the partial derivatives of the SSE with respect to all the weights and biases by the backpropagation algorithm.
(5) Update all weights and biases according to the gradient descent-based method.

After a long optimization run, the weights and biases gradually approach a specific limit and the decreasing of the SSE becomes more and more difficult. By now, it is the moment to stop training and recover the NPP according to Eq. (4.5). The residual between the MLP output and the target values can be evaluated and judged, and the MLP is ready for a real measurement test if the residual is small enough. If the residual is still unacceptably large, more neurons are exposed to the trial. If more neurons do not help, then the quality of the training data may need to be checked carefully because a high-quality database is the foundation of any data-driven method.

To use a trained MLP that is evaluated to be accurate enough, let $\mathbf{M}$ be an observed DI in a real measurement, which has a NFV denoted as $\boldsymbol{\omega}$. Then $\boldsymbol{\omega}$ is fed into the trained MLP to acquire the final output values. Next, the final output values are recovered with the means and standard deviations of the training database, and this product is the measured posture parameters. Finally, flowcharts for the MLP training procedure and the real measurement are depicted in Fig. 6.11 to summarize this section.

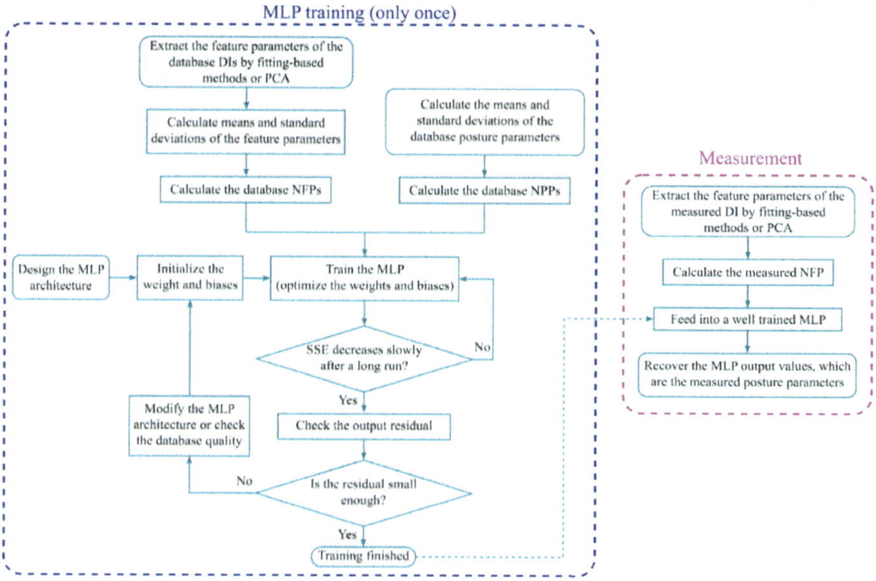

Fig. 6.11 Flowchart of the MLP training procedure and the real measurement

References

1. Rahlves M, Roth B, Reithmeier E (2015) Systematic errors on curved microstructures caused by aberrations in confocal surface metrology. Opt Express 23(8):9640–9648
2. Béguelin J et al (2020) Correction of spherical surface measurements by confocal microscopy. Meas Sci Technol 31(7):075002
3. Goodman JW (2005) Introduction to Fourier optics. Roberts & Co. Publishers, Englewood, Colorado
4. Schröder S et al (2011) Modeling of light scattering in different regimes of surface roughness. Opt Express 19(10):9820–9835
5. Schnars U, Jüptner WPO (2002) Digital recording and numerical reconstruction of holograms. Meas Sci Technol 13(9):R85
6. Pascanu R, Mikolov T, Bengio Y (2013) On the difficulty of training recurrent neural networks. In: International conference on machine learning. PMLR, 2013, pp 1310–1318
7. Glorot X, Bengio Y (2010) Understanding the difficulty of training deep feedforward neural networks. In: Proceedings of the thirteenth international conference on artificial intelligence and statistics. JMLR workshop and conference proceedings, 2010, pp 249–256
8. Wu G-W, Chen L-C (2023) Resolving measurement ambiguity in diffractive image microscopy for 6DOF surface measurement using designed aberration and multiple-layer perceptron. Int J Adv Manuf Technol. https://doi.org/10.1007/s00170-023-11147-0
9. Wu G-W, Chen L-C (2021) Precise 3-D microscopic profilometry using diffractive image microscopy and artificial neural network in single-exposure manner. Opt Lasers Eng 147:106732
10. Wu G-W, Nguyen DT, Chen L-C (2019) Novel full-field micro surface profilometry using diffractive image correlation without vertical scanning. Opt Lett 44(14):3534–3537
11. Wu G-W (2022) Theory and implementation of microscopic diffractive image profilometry. Ph.D. thesis, National Taiwan University

Chapter 7
Practical Implementation of Diffractive Image Profilometry (DIP)

7.1 Introduction

This chapter presents the overall system configuration in the first phase and provides detailed information on the components and design in the second phase. The third phase includes the design principle of a 3-D folded system. Finally, as preparation for the real measurement, the design of measuring arrays and the procedure of database building are explained.

7.2 System Design

Chapter 5 proposes a model to describe a pointwise DIP system. However, extending the concept to a multiple-point version for higher measurement speed in real applications is necessary. The digital micromirror device (DMD) is considered to replace the stationary illumination pinhole in Fig. 6.1. Furthermore, the actual system adopts an infinite conjugate configuration. As a result, the ideal system depicted in Fig. 6.1 is modified into Fig. 7.1 for the actual experiment setup.

In Fig. 7.1, light emitted from the lamp is directed by a light guide, which is then collected by a condenser to provide a converging illumination onto the DMD chip. Notice that the illumination direction onto the DMD chip is oblique in the 3-D space. However, for simplicity, an unfolded version is illustrated here, and details will be introduced in the following sections. Next, the DMD modulates the incoming light, for example, by three turned-on and spaced micromirrors. The modulated light beams are directed downward by the beam splitter and collimated by a tube lens. A customized aberration film is inserted between the tube lens and the objective to provide additional aberrations. The purpose of inserting this aberration film is to break the possible ambiguity mentioned in Chap. 5, which can occur when the system is “too perfect”.

L. Chen et al., *Diffractive Image Microscopy for 3D Imaging*, Springer Tracts in Mechanical Engineering, https://doi.org/10.1007/978-981-97-7782-2_7

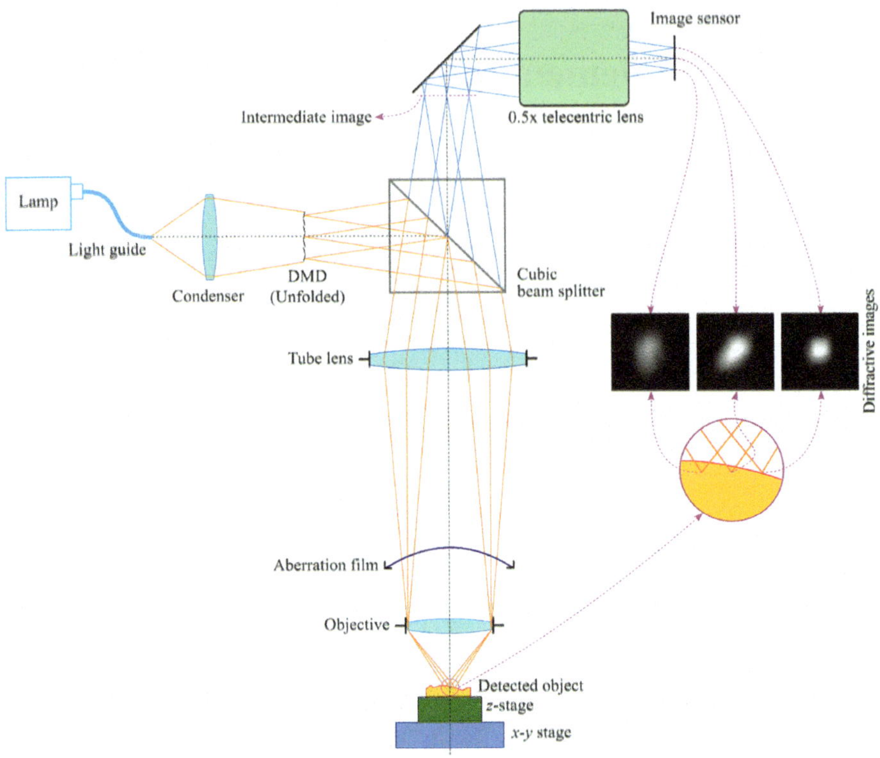

Fig. 7.1 The configuration of a diffractive image profilometer for integrating a DMD as a programmable pinhole array

After the light passes through the aberration film, it is focused on the detected object via an objective, simultaneously illuminating three distinct regions spaced apart. Consequently, the light reflected from these three regions is collected by the objective, passes through the aberration film again, and is focused via the tube lens to form an intermediate image. A 0.5× telecentric lens then shrinks the intermediate image to fit the size of the image sensor. Finally, three-spaced DIs are captured in a single image. After correctly processing these three DIs, posture parameters corresponding to the illuminated regions can be resolved. In addition to the optical system, a $x-y$ stage is employed for lateral scanning, and a z-stage is used to position the detected object within the measuring range.

It is important to note that in Fig. 7.1, a beam splitter is placed between the tube lens and the objective in the infinite conjugate system. Consequently, two tube lenses are required for the illumination and detection arms. While this can bring extra aberration to the system (especially spherical aberration and astigmatism), these are encoded in a built database, enabling the proposed method to remain effective.

7.3 Component and Design Details

Referring to Fig. 7.1, the Xenon lamp utilized is a Hayashi LA-410UV-3 (seen in Fig. 7.2a), offering a maximum illumination of 2,000,000 lx. Its spectrum is shown in Fig. 7.2b, where the visible light dominates. The light guide is a glass fiber bundle with a 4 mm exit diameter and NA 0.44.

The condenser used to collect the light emitted from the light guide is a plano-convex lens (PCL) (Edmund #45-246), which has a focal length of 75 mm (at 587.6 nm) and a clear aperture of φ49 mm. This PCL is placed about 50 mm from the end of the light guide to provide an illumination light cone with a 2.34° half converging angle onto the DMD. The DMD is placed about 150 mm away from the PCL. This converging angle ensures that the entry pupil of the downstream system is fully filled, aligning with the assumption made in theoretical simulations of the system.

The DMD is a micro-electro-mechanical system (MEMS). It comprises millions of micromirrors that can switch ON and OFF individually, as shown in Fig. 7.3a. As shown in Fig. 7.3b, when a micromirror is switched to the on-state, it reflects the incoming light toward a downstream optical system. Conversely, when switched to the off state, the micromirror reflects the incoming light to a beam dump, absorbing the reflected light. The adopted DMD is the TI DLP9000 evaluation module (Fig. 7.4). It has 2560 × 1600 micromirrors with 7.56 μm pitch. The chip size is about 19.35 × 12.10 mm^2 with a diagonal length of 22.82 mm. Each micromirror can tilt up to + 12° or down to − 12° along the diagonal as shown in Fig. 7.4. Consequently, the illumination chief ray to the DMD should align with the direction depicted in the left subfigure in Fig. 7.4. If the micromirrors are switched ON and OFF without specific pulse width modulation (PWM), the highest switching speed is 9523 Hz, which is considerably fast when compared to a typical image sensor. The DMD enables a wide range of applications to be realized.

Right after the DMD, the beam splitter is a 45 mm 50/50 cube beam splitter (Edmund #32-703). The tube lens beneath the beam splitter is the Thorlabs TTL200MP, featuring a focal length of 200 mm, a f-number of 10 (~NA 0.05),

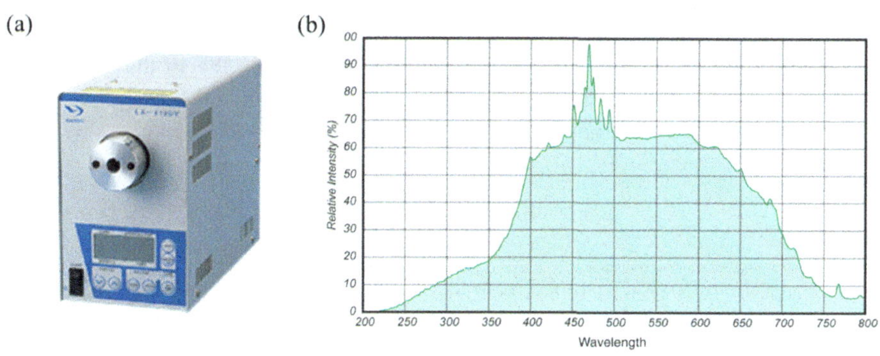

Fig. 7.2 **a** The used Xenon lamp and **b** its spectrum [1]

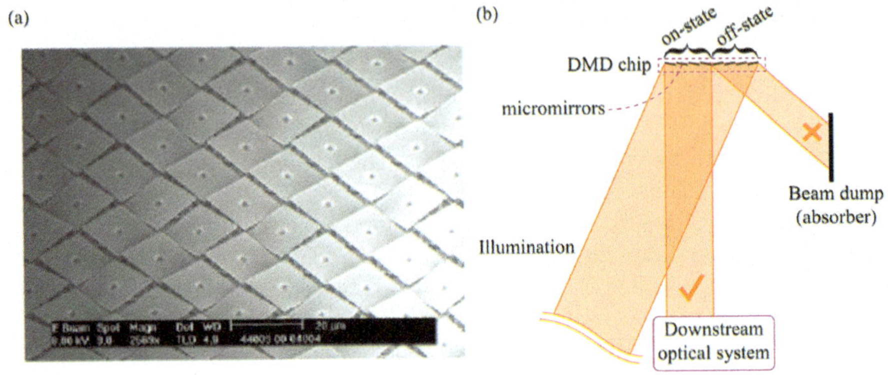

Fig. 7.3 **a** A scanning electron microscopy (SEM) image of a DMD [2]. **b** The light modulation mechanism of a DMD. Detailed diffraction phenomenon is ignored

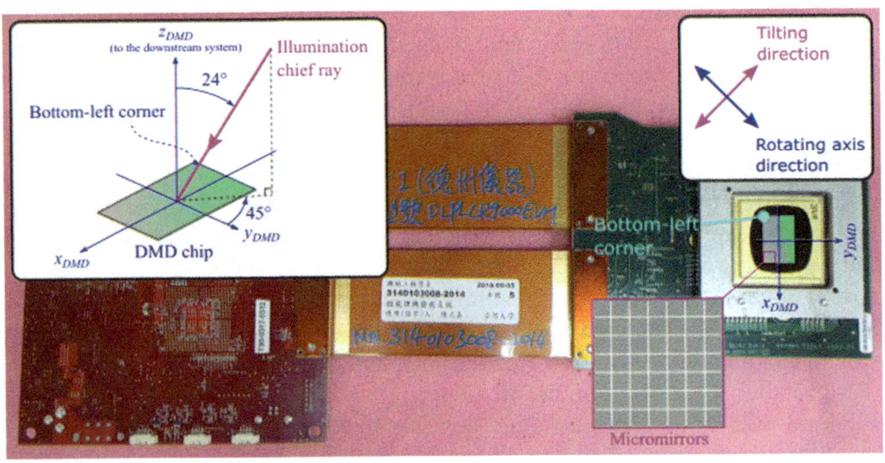

Fig. 7.4 TI DLP9000 DMD module

and a field size of $\varphi 22$ mm. This field size corresponds to 95.9% of the DMD diagonal, suggesting that while imaging quality at the corner of the DMD may be slightly worse, it is not noticeably apparent. The position of this tube lens is given by

$$d_{\mathrm{dt}} = d_{\mathrm{twd}} + t_{\mathrm{bs}}\left(1 - \frac{1}{n(\lambda_{\mathrm{ref}})}\right), \tag{7.1}$$

where d_{dt} is the unfolded distance from the DMD to the tube lens. d_{twd} is the working distance of the tube lens, which is 148 mm. t_{bs} is the thickness of the beam splitter, which is 45 mm. $n(\lambda_{\mathrm{ref}})$ is the refractive index at the reference wavelength λ_{ref}. Let

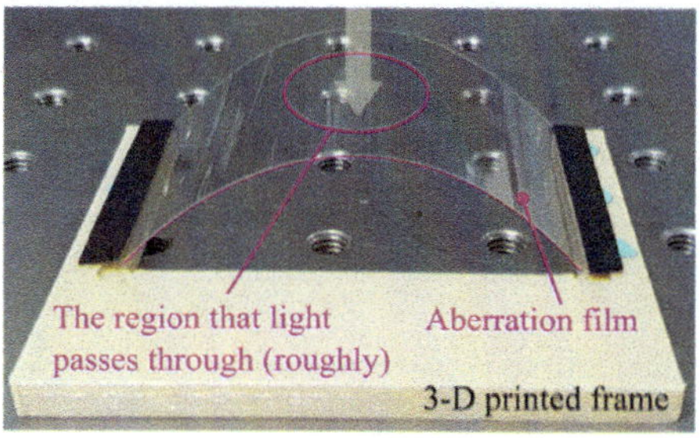

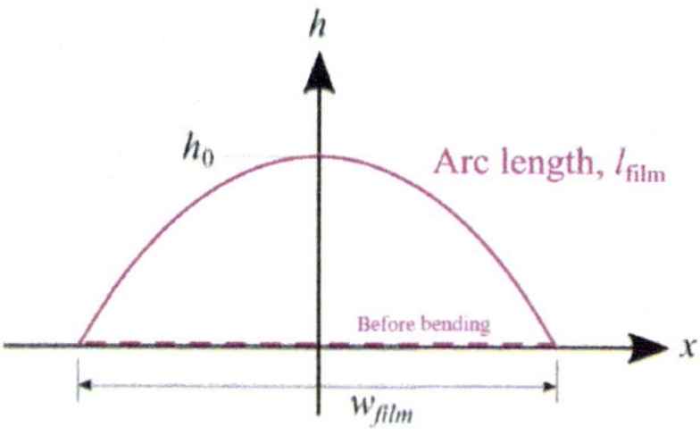

Fig. 7.5 **a** The aberration film. **b** a parabolic model that describes the shape of the aberration film

λ_{ref} be 0.55 μm, $n(\lambda_{ref})$ for the beam splitter (made of N-BK7) is 1.5185. With these conditions, d_{dt} is calculated to be 163.365 mm.

The customized aberration film is a bent projector film (3 M, PP2910, 100 μm, polypropylene film), trimmed and bent, then fixed in a 3-D printed frame as shown in Fig. 7.5a. However, if numerical simulation of the aberration film is desired, its geometry information becomes necessary. Assuming a parabolic approximation for the shape of the bent film, its configuration is determined by the "width" after bending, w_{film}, and its original length l_{film}, which is assumed to equal the arc length. Now consider the parabolic equation that describes the film shape as

$$h(x) = h_0 - \frac{4h_0}{w_{film}^2}x^2, \tag{7.2}$$

where h_0 is the center height and x is defined in Fig. 7.5b. Then, the arc length is given by

$$\begin{aligned} l_{film} &= \int_{\frac{1}{2}w_{film}}^{-\frac{1}{2}w_{film}} \sqrt{1+\left(\frac{dh}{dx}\right)^2}\,dx \\ &= \frac{w_{film}}{2}\sqrt{1+\frac{16h_0^2}{w_{film}^2}} + \frac{w_{film}^2}{16h_0}\ln\left|\frac{4h_0+\sqrt{w_{film}^2+16h_0^2}}{-4h_0+\sqrt{w_{film}^2+16h_0^2}}\right|, \end{aligned} \tag{7.3}$$

where w_{film} is determined by the mechanical design of the 3-D printed frame. l_{film} is the original length of the aberration film, which can be trimmed arbitrarily according to the need. The larger the extra aberration required, the larger l_{film} needs to be. With predetermined w_{film} and l_{film}, the only unknown h_0 can be numerically solved by Newton's root-finding method. Figure 7.6 shows the relationships between l_{film} and h_0 when w_{film} is 60 or 70 mm.

With explicit values for w_{film} and h_0, the shape of an arbitrary aberration film can be described in commercial optical design software, enabling numerical simulation.

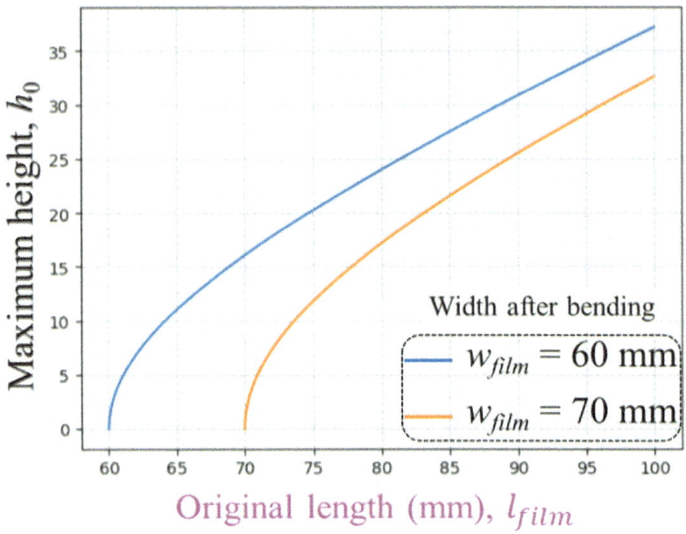

Fig. 7.6 The relationships between the maximum height and the original length of the aberration film. Two examples are given

The resolved parabolic equation is considered the top surface of the aberration film, and the bottom surface is acquired by shifting the top surface downward according to the thickness of the film. With the above-proposed method, the shape of an arbitrary aberration film can be modeled, and its contributed aberration can be simulated in software.

For example, in Fig. 7.7a, collimated light beams enter from the left-hand side, and a 100 μm thick aberration film with $l_{\text{film}} = 62$ mm and $w_{\text{film}} = 60$ mm is placed in front of an ideal lens. Apparent astigmatism can be observed through a focus spot diagram, as shown in Fig. 7.7b. More quantitatively, we can check the first 10 Zernike coefficients, where the defocus term (Z_2^0, ~ 0.017) and the vertical astigmatism term (Z_2^2, ~ − 0.024) are the only two nonzero terms. The defocus term is also visible in the center column of Fig. 7.7b, although it is not particularly severe. This method provides a semi-quantitative approach to designing an aberration film, minimizing the need for extensive trial and error to achieve a "just fine" level of aberration. In this context, a "just fine" aberration means that the aberration is enough to break the ambiguity without being unnecessarily large. However, there may not be a definitive criterion to judge a "just fine" aberration. Here, as a reference, a quantitative criterion that defines a "just fine" aberration is suggested as

$$\frac{\sigma_{u\text{ref}}}{\sigma_{v\text{ref}}} \geq c_{\min}, \tag{7.4}$$

where $\sigma_{u\text{ref}}$ and $\sigma_{v\text{ref}}$ are the fitted standard deviations (Eq. 6.10) along the long-axis and short-axis defined in Sect. 6.5.2 when the defocus value, d_{f}, is η_{ref} times of the

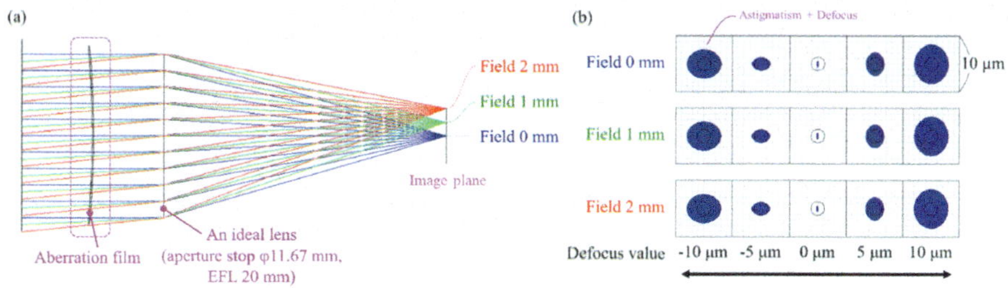

Fig. 7.7 **a** A simple setup that demonstrates the influence of the aberration film. **b** The through-focus spot diagram

objective's depth of focus (DOF). Where $c_{\min}$ and η_{ref} are two manually defined values, which may be chosen as 1.1 and 8.0, respectively. The DOF of the objective is defined as

$$\text{DOF} = \frac{\lambda}{2(\text{NA})^2} \tag{7.5}$$

with a wavelength of λ and a numerical aperture of NA.

Beneath the aberration film, a 10× objective (Mitutoyo 378-803-3, NA0.28, DOF 3.5 μm) is used to illuminate the detected object and collect the reflected signal light. The tube lens's specifications determine the distance from the tube lens to the objective, which should be 228 mm from the flange of the tube lens to the flange of the objective.

After the signal light travels upwards and forms an intermediate image, its position corresponds to a conjugate position of the DMD. Therefore, the distance from the top surface of the tube lens to the intermediate image is also 163.365 mm, as determined by prior calculations.

Next, the intermediate image is reduced in size by a 0.5× telecentric lens (HI KEEN TXF-5MDT05X65-1C, NA 0.042), projecting onto the image sensor. The position of the telecentric lens is determined by its working distance, which is 65 mm. Finally, a camera (Basler acA2040-180km) served as the image sensor, featuring a pixel size of 5.5 μm and a resolution of 2048 × 2040 pixels. It has a dynamic range of 60 dB and can achieve a maximum framerate of up to 180 fps (full-frame).

The $x-y$ stage is a customized stage equipped with linear optical scales, providing a minimum step size of 0.1 μm. Additionally, the z-stage is a piezoelectric transducer module (PI P-725.4CD, Fig. 7.8) with a 400 μm travel.

Fig. 7.8 The used piezoelectric transducer module (PI P-725.4CD)

7.4 Integrated System

The system shown in Fig. 7.9a is introduced in the previous section, which is used for most experiments. However, in the early stage of this work, the DIP system aimed to integrate into a wafer stage, resulting in the configuration shown in Fig. 7.9b. For this purpose, the DIP system must be designed within a space limited by the granite structure in Fig. 7.9b. Therefore, the whole system is further designed into a compact 3-D folded module, with design principles outlined in the following.

It is required to properly design such a 3-D folded module for the highest standard, which is an enormous task. Essential concerns are classified and listed below:

(1) Mechanical and assemble accuracy

   i. Material.

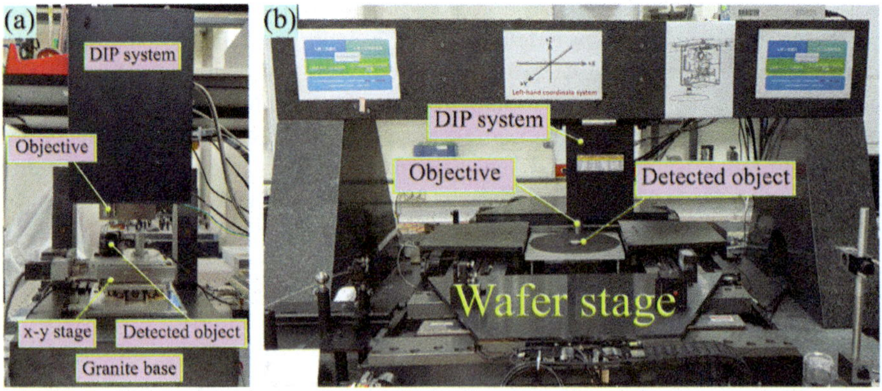

Fig. 7.9 **a** The final system used for most of the experiments. **b** An integrated system in the early stage of this work, in which the DIP system is integrated into a wafer stage

ii. Machining method.
iii. Positioning design.
iv. Tolerance design.
v. Opened degree of freedom for fine adjustment.

(2) Structure properties

i. Static properties, primarily weights, and gravitational deformation.
ii. Dynamic properties, the frequency response.

(3) Thermal properties

i. Heat dissipation design.
ii. Thermal deformation.

(4) Wiring
i. It is better to arrange all wires (light and electric wires) to the same side of the module to keep the appearance as neat as possible.

Other factors of concern are budget, the ability of the machining factory, and time. These considerations lead to the following practical design guidelines.

(1) All mechanical parts are made of aluminum alloy (A6061T6) for lightweight purposes.
(2) All mechanical parts that are mutually assembled must be designed with positioning slots and keys.
(3) All slots and keys are designed to have clearance fit within the range of 0.1–0.2 mm.
(4) The tolerance of important angles that may affect the direction of light beams is $\pm$ 0.05°.
(5) All mechanical parts are surface treated by "black matte" anodizing to minimize the influence of stray light.
(6) All positioning keys are "pressed" in the same direction against the positioning slots during the assembly process to minimize angular assembly error.
(7) The DMD and the camera are the two primary heat sources. DMD is fixed on the top board of the module with an attached heat sink exposed outside. The camera is placed "outside" of the module housing and attached to a heat sink fin. Therefore, both can efficiently dissipate the heat to the ambient environment.
(8) All electric wires and the light guide are arranged to the same side of the module for a neat appearance.

Following the above guideline, the 3-D folded module is designed as shown in Fig. 7.10a, b. Furthermore, the illumination and detection light paths are depicted in Fig. 7.11a, b, respectively.

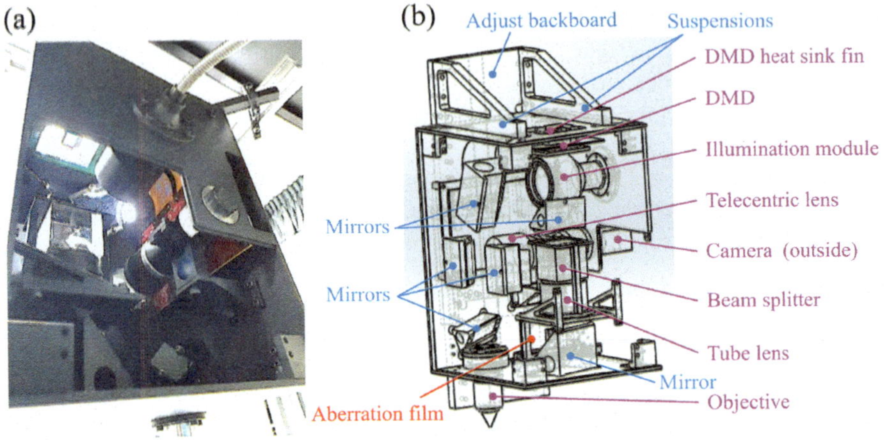

Fig. 7.10 **a** A photo of the 3-D folded DIP system and **b** the CAD mechanical design

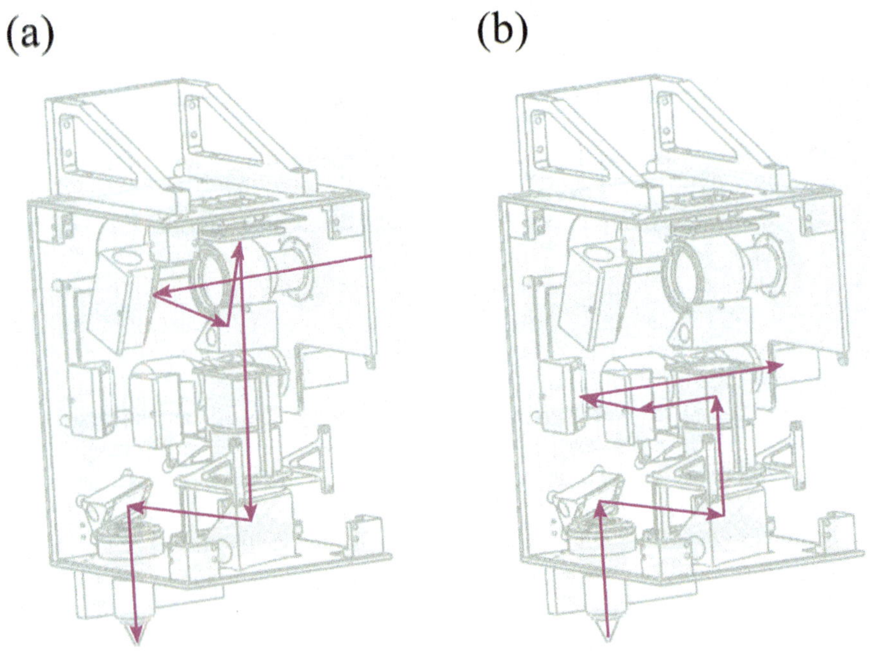

Fig. 7.11 **a** The illumination light path. **b** The detection light path

7.5 Measuring Array

Although the DMD can facilitate lateral parallelization, careful design, and definition of a measuring array are essential. For example, Fig. 7.12 shows a portion of a measuring array where a "pinhole" consists of a group of adjacent on-state micromirrors restricted to form a square for simplicity. Therefore, its size can be defined by the width of the square, denoted as p_s, which is 3 in the example shown in Fig. 7.12. Moreover, the pitch between adjacent pinholes is denoted as p_p, which is 8 in this example. Notice that p_p must be greater than p_s. Otherwise, the measuring array is effectively the same as turning on all the micromirrors. Next, as shown in Fig. 7.13, the whole measuring array will be shifted in a zig-zag matter to measure positions between pinholes. Finally, the scanning pitch is denoted as s_p. s_p must be a factor of p_p, excluding itself. In this example, p_p is 8, so the valid s_p can be 1, 2, and 4 micromirrors. In Fig. 7.13, the case of $s_p = 4$ is illustrated. Conclude from the above, p_s, p_p, and s_p together define a "scanning plan". The scanning plan process involves scanning an object and taking measurements within a single FOV. Furthermore, the steps to complete the scanning within a FOV can be calculated as $(p_p/s_p)^2$.

In addition to the above fundamental mathematical relationships, there are other important concerns. For example, the pinhole size should be minimized if the light source is sufficiently strong. A larger pinhole size results in a larger illumination region on the detected object, causing the measurement to represent an average result over the region. Second, the scanning pitch should be tailored to the specific application. A smaller scanning pitch indicates denser measuring points and longer measurement times, which may not always be necessary for all scenarios. Third, while a smaller pinhole pitch seems to provide a higher parallelization, it can limit the axial measurable range. When a detected object deviates from the ideal focal position, the image sensor observes a blurred "DI array". If the blur is too severe

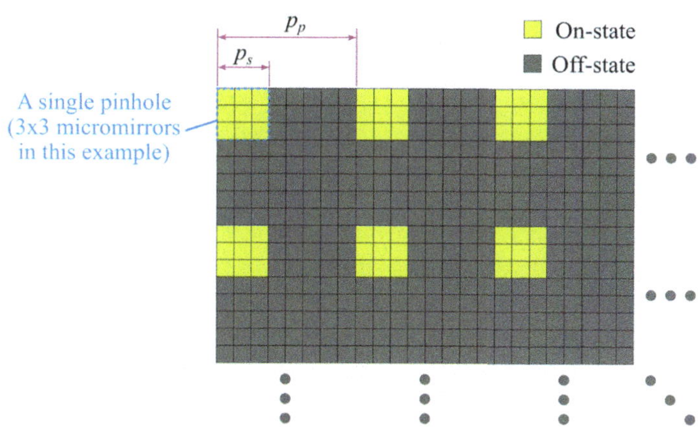

Fig. 7.12 The definition of pinhole, size, and pitch in a measuring array

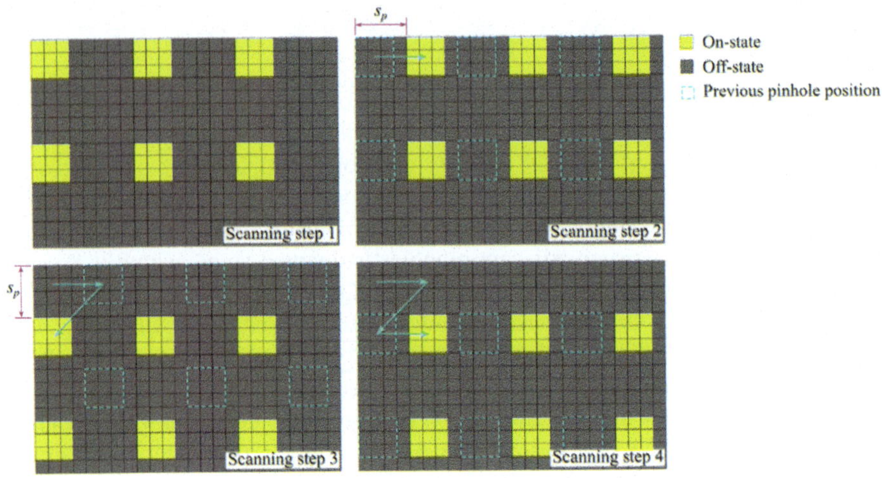

Fig. 7.13 The definition of scanning pitch and the zig-zag scanning

or the pinhole pitch is too small, adjacent DIs may overlap, leading to unwanted crosstalk. This phenomenon requires quantitative investigation, as outlined below.

Let's reconsider Fig. 7.1 and assume a pinhole on the DMD chip. If the pinhole has a size of p_s and the micromirror pitch is m_p, it can be considered a small square light source with a width of $p_s m_p$. The size of the DI corresponding to this pinhole can be estimated by

$$R_{DI} \cong \eta_{safe}\sqrt{R_{geo}^2 + R_{dif}^2}, \tag{7.6}$$

where R_{DI} is the radius of the DI, R_{geo}, and R_{dif} are the two terms corresponding to the contribution of geometrical and wave optics. η_{safe} is a safety factor. The geometrical optics term is estimated by

$$R_{geo} \cong p_s m_p M_{di} + 2 d_f M_{oi}^2 \mathrm{NA}_{img}, \tag{7.7}$$

where M_{di} is the transverse magnification from the DMD to the image sensor. The first term refers to the geometry projection size of the pinhole onto the image sensor. The second term accounts for defocus on the object side, where d_f is the defocus value of the objective side, M_{oi} denotes the transverse magnification from the object side to the image sensor side, and NA_{img} is the numerical aperture of the image sensor side. Finally, the coefficient 2 arises from the assumption of reflection from the object.

Next, the wave optics term can be estimated by the Airy disk radius at the image sensor side, which is given by

$$R_{\text{dif}} \cong \frac{0.61\lambda_{\text{ref}}}{\text{NA}_{\text{img}}}, \tag{7.8}$$

at a reference wavelength λ_{ref} (e.g., 0.55 μm). Finally, to avoid overlapping of two adjacent DIs on the image sensor, it must satisfy

$$2R_{\text{DI}} < p_{\text{p}} m_{\text{p}} M_{\text{di}}. \tag{7.9}$$

Equation (7.5)–(7.8) give a restriction among p_{s}, p_{p}, and d_{f}. For example, Fig. 7.14 shows the relationship between the maximum d_{f} and p_{p}, with the conditions listed in Table 7.1 The only difference between these two conditions is that the pinhole size in condition 1 is smaller. Consequently, the maximum defocus value without overlapping DIs can be larger than condition 2. Figure 7.14 can also be interpreted conversely. For instance, if the desired defocus range of a database is between ± 30 μm, then the pinhole pitch should be at least 43 or 48 micromirrors for condition 1 and condition 2, respectively.

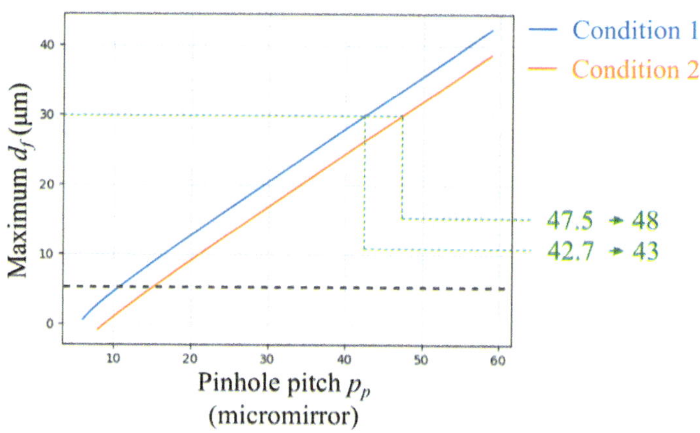

Fig. 7.14 The relationship between the maximum defocus value and the pinhole pitch without overlapping DIs. Detailed conditions are listed in Table 7.1

Table 7.1 Detailed conditions for Fig. 7.14

Items (unit)	Condition 1	Condition 2
m_{p} (μm)	7.56	7.56
p_{s} (micromirror)	1	3
M_{di} (−)	0.5	0.5
M_{oi} (−)	5	5
λ_{ref} (μm)	0.55	0.55
NA_{img} (−)	0.042	0.042
η_{safe}	1.2	1.2

Besides the above restrictions, as the defocus value increases, the average intensity of the DI decreases. However, the background noise becomes relatively significant if the average intensity is too low. Thus, it is more challenging to extract useful information from the DI. In other words, this is a signal-to-noise ratio (SNR) issue, with its limit determined experimentally.

7.6 Preparation for Database Building

A database building module (DBM) shown in Fig. 7.15 is employed to collect DIs corresponding to various posture parameters. This module comprises a 2-D beam steering mirror (BSM, Optotune, MR-E-2, resolution < 5 μrad, Fig. 7.16), a z-stage, and a $x - y$ stage. The BSM incorporates a built-in mirror capable of rotation along two orthogonal axes. However, before using this DBM, a specialized and careful alignment process between the DBM and DIP system is necessary.

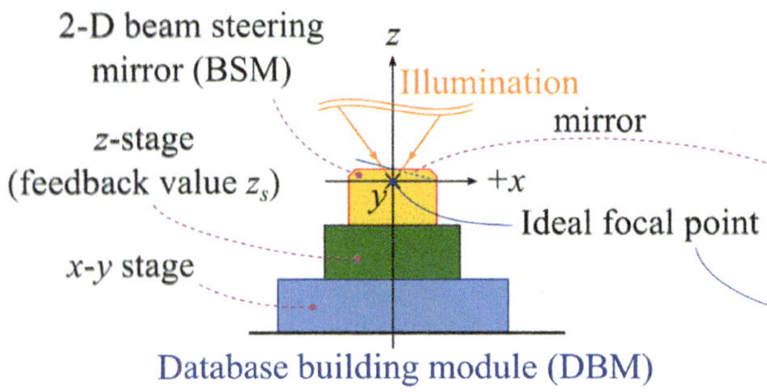

Fig. 7.15 A database building module comprises a 2-D beam steering mirror, a z-stage, and a $x - y$ stage. It can generate desired surface orientations

Fig. 7.16 The used beam steering mirror (Optotune, MR-E-2)

7.6.1 Align the Optical Axis and the z-Axis

The first alignment goal is to ensure the optical axis is parallel to the z-axis of the stage since all components will be placed on it. To achieve this, the DMD is set up to open a pinhole for observation. Next, a flat optical mirror is placed on the stage and adjusted along the x and y directions. If the optical axis is not parallel to the z-axis, the observed DI can defocus as the stage moves. The degree of defocusing can be determined by adjusting the z-stage to identify a position where the DI is well-focused. Therefore, the angle between the optical axis and the z-axis can be calculated by the displacement along the x or y directions and the degree of defocusing. With this information, the relative orientation of the DIP system and the stage can be adjusted accordingly. As the cornerstone of the experiment, the setup system is carefully adjusted to ensure that the change in defocus value is within 1 μm, while the traveling distances along the x or y directions are 5 mm. This translates to an angle between the optical axis and the z-axis of approximately 0.012° or less.

7.6.2 Align the Optical Axis with the Normal Vector of the DBM

The second alignment goal is to ensure the parallelism between the optical axis and the normal vector of the DBM when it is set to be horizontal. Therefore, the DBM is first placed under a DIP system, with the tilting angle set to $\theta_{\text{tilt}} = 0$. At the same time, the DMD is configured to project a reference illumination array onto the mirror. Subsequently, an image resembling Fig. 7.17a might be observed via the image sensor. This image reveals that the mirror of the DBM is not perpendicular to the optical axis because the degree of focusing is not uniform across the FOV, despite the tilting angle being set to zero.

A quantitative description of the degree of alignment should be defined to minimize misalignment. It can be done by quantifying the size of "spots" in Fig. 7.17, with each spot being localized based on the predesigned DMD pattern. Subsequently, a region of interest (ROI) is cropped, as shown in the subfigure in Fig. 7.17a. Within this ROI, the spot's centroid is calculated according to

$$\begin{cases} m_c = \left(\sum\limits_{m=1}^{M}\sum\limits_{n=1}^{N} g(m,n)m\right) \Big/ \left(\sum\limits_{m=1}^{M}\sum\limits_{n=1}^{N} g(m,n)\right) \\ n_c = \left(\sum\limits_{m=1}^{M}\sum\limits_{n=1}^{N} g(m,n)n\right) \Big/ \left(\sum\limits_{m=1}^{M}\sum\limits_{n=1}^{N} g(m,n)\right) \end{cases}, \tag{7.10}$$

where m_c and n_c are the row and column coordinates within the ROI of the centroid, respectively. m and n are the row and column indices. $g(m, n)$ is the grayscale value (GV) at (m, n) pixel within the ROI. The "spot radius" denoted by r_{sp} can be defined as

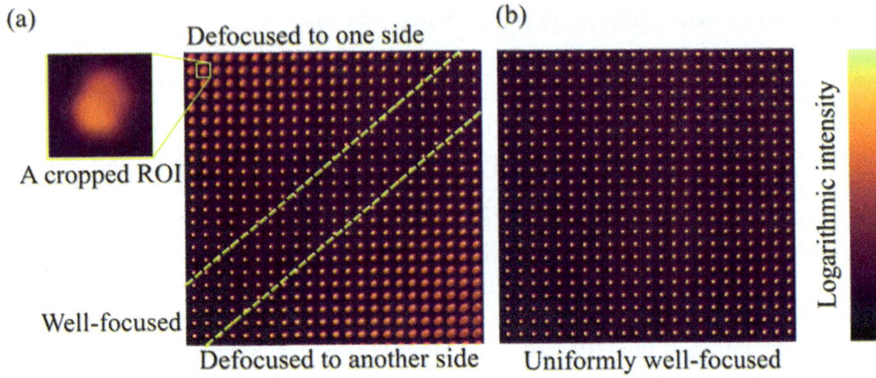

Fig. 7.17 Observed images when aligning the optical axis with the normal vector of the DBM. **a** Observed image when the mirror of the DBM is not perpendicular to the optical axis. **b** The observed image after the DBM is well-adjusted

$$r_{sp} = \frac{\sum_{m=1}^{M}\sum_{n=1}^{N} g(m,n)\sqrt{(m-m_c)^2+(n-n_c)^2}}{\sum_{m=1}^{M}\sum_{n=1}^{N} g(m,n)}. \tag{7.11}$$

The physical meaning of Eq. (7.10) is the GV weighted average distance with respect to the centroid of the spot for all pixels. With this definition, the above alignment problem becomes a minimization problem, aiming to minimize the sum of all spot sizes across the FOV. For example, the summation of all spot sizes in Fig. 7.17a is 3478.69 pixels. After minimization, the summation of all spot sizes in Fig. 7.17b drops to 2879.08 pixels. Following this concept, a convenient software tool capable of calculating the instantaneous sum of all spot sizes in real time will be helpful for manual alignment processes. Automation of this process is achievable if controllable actuators are integrated with the setup system. Following the above alignment steps, the optical axis is considered perpendicular to the mirror of the DBM.

7.6.3 *Align the Transverse Axis with the Rotation Axes of the DBM*

The following alignment goal is to align the two rotation axes so that they are parallel to the x-axis and y-axis. Here, the x-axis is defined by the row direction of the DMD, and the y-axis is defined by the column direction of the DMD. It is important to note that at this stage, the origin of the $x-y$ plane is not yet defined; only their directions are defined. While it seems to be the intersection of the $x-y$ plane and the optical axis in Fig. and Fig. 7.15, it is defined by a specific field point of interest, which will be explained in the next alignment step.

To achieve this goal, the BSM is set as $\theta_{\text{tilt}} = \theta_{\text{test}}$ and $\phi_{\text{tilt}} = 0$. If $\theta_{\text{test}} \neq 0$ (e.g., $\theta_{\text{test}} = 5°$ for an NA 0.28 objective), an image like Fig. 7.18a might be observed, wherein a virtual line with the smallest spot size appears. This virtual line denotes one of the rotation axes of the DBM, and it can be acquired by fitting the spot radius distribution with a second-order polynomial surface defined by

$$\begin{cases} r_{\text{sp}}(m'_{\text{fov}}, n'_{\text{fov}}) = c_0 + c_1 n'_{\text{fov}} + c_2 n'^2_{\text{fov}} \\ \begin{bmatrix} m'_{\text{fov}} \\ n'_{\text{fov}} \end{bmatrix} = \begin{bmatrix} \cos\theta_{\text{msp}} & -\sin\theta_{\text{msp}} \\ \sin\theta_{\text{msp}} & \cos\theta_{\text{msp}} \end{bmatrix} \begin{bmatrix} m_{\text{fov}} \\ n_{\text{fov}} \end{bmatrix} \end{cases}, \tag{7.12}$$

Referring to Fig. 7.18, where m_{fov} and n_{fov} are the row coordinate and column coordinate of the FOV, respectively. m'_{fov} and n'_{fov} are the row and column indices of the rotated coordinate. $c_0, c_1, c_2, \theta_{\text{msp}}$ are fitting coefficients. θ_{msp} is the angle from the m_{fov}-axis to the m'_{fov}-axis, which is of particular interest for this alignment purpose. A quantitative description of this alignment process involves rotating the whole DBM to minimize θ^2_{msp}. The square serves to facilitate a smoother optimization process around the best point. For example, Fig. 7.18a shows the original observed image. Its spot radius distribution is shown in Fig. 7.18b, and the corresponding fitting result is shown in Fig. 7.18c. The result indicates that the virtual line is positioned at a certain angle from the edge of the image. After rotating the whole DBM accordingly, the

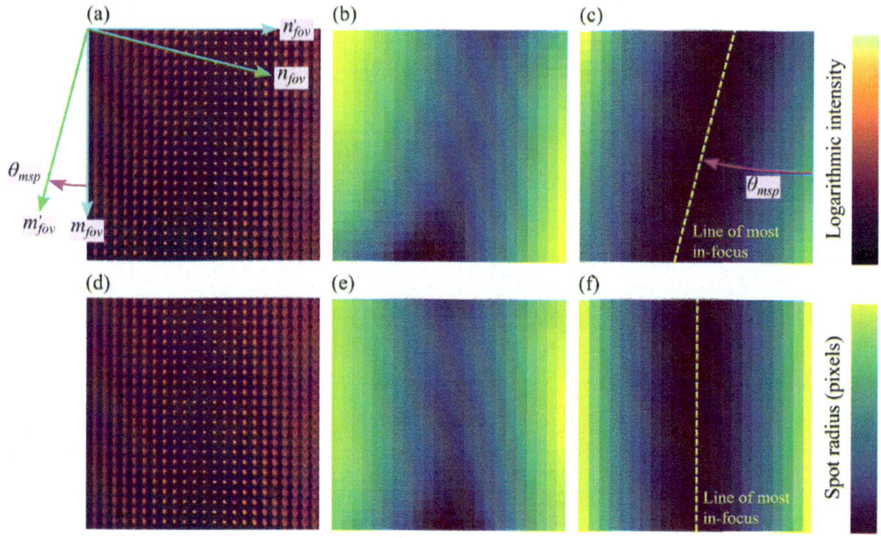

Fig. 7.18 An illustration of the alignment process described in Sect. 7.6.3: **a** Observed image when the rotation axes are not aligned yet. **b** The original spot radius distribution. **c** The original fitted spot radius distribution. **d** Newly observed image after aligning the rotation axes. **e** The new spot radius distribution. **f** The new fitted spot radius distribution

newly observed image is shown in Fig. 7.18d. Figure 7.18e, f show the corresponding spot radius distribution and the fitting result. After this alignment, the rotation axes of the DBM are considered to be parallel to the x-axis and y-axis.

7.6.4 Align the Rotational Axes of the DBM with the Field Point of Interest

In line with the summary in Sect. 6.3, a database depends on the pinhole position, width, and the associated cropping region. Suppose these three factors are established, and the selected region in Fig. 7.19a is the corresponding cropping region, the fourth alignment goal is to ensure that "two" virtual lines pass through the center of this region. The first virtual line is hidden in Eq. (7.11), which can be explicitly written as

$$L_{\mathrm{ay}} : n_{\mathrm{fov}} + m_{\mathrm{fov}} \tan\theta_{\mathrm{msp}} + \frac{c_1}{2c_2 \cos\theta_{\mathrm{msp}}} = 0. \tag{7.13}$$

If the DBM is well aligned, i.e., $\theta_{\mathrm{msp}} \cong 0$, Eq. (7.12) reduces to

$$L_{\mathrm{ay}} : n_{\mathrm{fov}} + \frac{c_1}{2c_2} = 0, \tag{7.14}$$

which is a line that is parallel to the y-axis, as shown in Fig. 7.19b. The second virtual line can also be acquired using the same approach, but with setting $\theta_{\mathrm{tilt}} = \theta_{\mathrm{test}}$ and $\phi_{\mathrm{tilt}} = 90°$. This time the fitted θ_{msp} will be about 90°, then Eq. (7.12) reduces to

$$L_{\mathrm{ax}} : m_{\mathrm{fov}} + \frac{c_1}{2c_2} = 0, \tag{7.15}$$

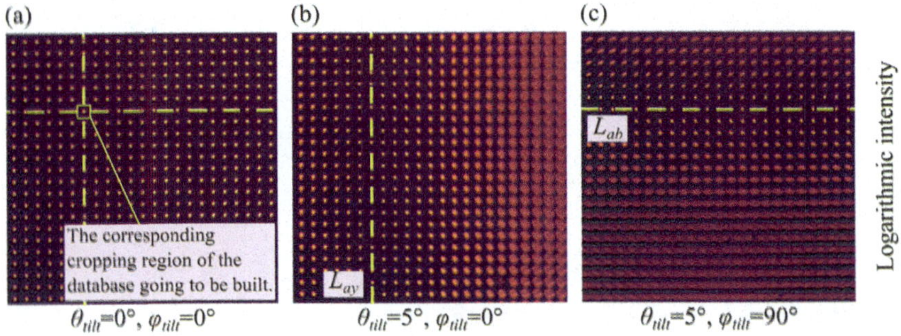

Fig. 7.19 An illustration of the alignment of the stationary point of the BSM with the center of the cropping region

which is a line that is parallel to the x-axis, as shown in Fig. 7.19c. The horizontal position of L_{ay} can be adjusted by moving the DBM along the x-direction. Similarly, the vertical position of L_{ax} can be adjusted by moving the DBM along the y-direction. Therefore, by moving the $x - y$ stage, the intersection of L_{ay} and L_{ax} can coincide with the center of the selected region in Fig. 7.19a. When this is achieved, the z position of the BSM surface within the selected region remains nearly unchanged with adjustments in θ_{tilt} and ϕ_{tilt}. This is because L_{ay} and L_{ax} are the two rotation axes of the BSM. Thus, their intersection is a stationary point fixed at a certain point in the 3-D space, regardless of alterations in θ_{tilt} and ϕ_{tilt} (except for mechanical error).

Like the previous two alignment steps, a quantitative expression for this alignment process can be defined by

$$\left(\hat{n}_{\text{fov}} + \frac{c_1}{2c_2}\right)^2 + \left(\hat{m}_{\text{fov}} + \frac{c_1}{2c_2}\right)^2. \tag{7.16}$$

This equation is derived from the summation of the squared residuals of L_{ay} and L_{ax}, where $(\hat{m}_{\text{fov}}, \hat{n}_{\text{fov}})$ is the coordinate of the center of the selected region within the FOV. Nevertheless, this alignment process does not require extreme precision because its purpose is to prevent severe defocusing of DIs within the selected region when adjusting the orientation of the BSM. It can be done manually without precise image processing and calculation. After completing all four alignment steps, the system is ready for database building.

7.7 Sampling Plan

To build a database with various posture parameters, the sampling range and pitch (step size) of d_{f}, θ_{tilt}, and ϕ_{tilt} are need to be decided. This process is referred to as a sampling plan. A denser sampling preserves more detailed information about the variation but can result in impractical time requirements for data collection and subsequent usage. Conversely, a database with insufficient sampling density may compromise accuracy. Therefore, devising a reasonable and manageable sampling plan is crucial.

7.7.1 Sampling Plan for Surface Height

The primary concern of the sampling range of d_{f} is the SNR, which can be defined by

$$\text{SNR} = \frac{\text{GV}_{\text{avgmax}}}{\text{GV}_{\text{bn}}}, \tag{7.17}$$

where GV_{avgmax} is the average maximum GV of multiple captured DIs and GV_{bn} is the average background noise. When d_f becomes larger, the SNR become smaller. If the SNR is below a certain limit, the information hidden in a DI cannot be effectively extracted. Such a limit is typically determined experimentally.

The used objective's DOF can be a meaningful reference for determining the sampling pitch. It is suggested that the sampling pitch of d_f should be smaller than half of the objective's DOF.

7.7.2 *Sampling Plan for Tilting Angle*

Similarly, the objective's NA is a convincing reference for establishing the sampling range of the tilting angle, as it determines the maximum angle at which the reflected light can be collected. Therefore, the sampling range of the tilting angle is chosen from zero (horizontal state) to the angle defined by

$$\theta_{tilt,max} = \sin^{-1}(NA). \tag{7.18}$$

The sampling pitch can be chosen to be an arbitrary small angle smaller than one-tenth of $\theta_{tilt,max}$, for example 1°.

7.7.3 *Sampling Plan for Tilting Direction*

The sampling range of the tilting direction is 0–360°. As for the sampling pitch, because the tilting direction is of lesser importance, it can be larger to save the database building time, for instance, 10°.

7.7.4 *Number of DIs in a Database and a Sampling Plan Example*

With the established sampling plan, the total number of DIs within one database can be calculated according to

$$N_{DI} = \left\lfloor \frac{R_{df}}{P_{df}} + 1 \right\rfloor \left\lfloor \frac{R_{ta}}{P_{ta}} \right\rfloor \left\lfloor \frac{R_{td}}{P_{td}} \right\rfloor + \left\lfloor \frac{R_{df}}{P_{df}} + 1 \right\rfloor, \tag{7.19}$$

where N_{DI} is the total number of DIs within the database. R_{df} and P_{df} are the sampling range and pitch for d_f. R_{ta} and P_{ta} are the sampling range and pitch for θ_{tilt}. R_{td} and P_{td} are the sampling range and pitch for ϕ_{tilt}. Furthermore, the notation 

Table 7.2 An example plan for posture parameters sampling when assuming the objective's NA is 0.28

Posture parameters (unit)	Sampling range	Sampling pitch
d_f (μm)	60 (− 30 to + 30)	1
θ_tilt (°)	8° (0° to 8°)	1°
ϕ_tilt (°)	360° (0° to 360°)	10°

 is the floor function, which rounds down to the largest integer not greater than the value inside. For example, $\lfloor 1.23 \rfloor = 1$ and $\lfloor 5.0 \rfloor = 5$. In Eq. (7.18), the inconsistent expression over the three parameters is due to the following two facts:

(1) When $\theta_\mathrm{tilt} = 0$, the BSM is just a horizontal mirror. At this point, any ϕ_tilt leads to the same DI.
(2) Tilting direction, ϕ_tilt, is a cyclic parameter, i.e., the case of $\phi_\mathrm{tilt} = 0°$ is the same as the case of $\phi_\mathrm{tilt} = 360°$.

Based on Eq. (7.18), the required database building time can be estimated accordingly. Finally, a sampling plan example is given in Table 7.2 as an easy-understanding summary. The total number of DIs in this particular example is 17,629.

7.8 Database Building

After careful alignment and decision of a sampling plan, the DBM and DIP system can be controlled by a program to collect DIs corresponding to all specified posture parameters automatically. These DIs are denoted as $\mathbf{D}_{i,j,k}$, corresponding to the i-th z_s, $z_{\mathrm{s},i}$, j-th θ_tilt, $\theta_{\mathrm{tilt},j}$, and k-th ϕ_tilt, $\phi_{\mathrm{tilt},k}$. Note that in this phase, the program only knows the feedback value of the z-stage, z_s. The desired axial position defined with respect to the ideal focal point, d_f, is not yet known. Furthermore, notably, the relationship between z_s and d_f is not simply differed by a constant. For example, consider Fig. 7.20, where the rotation axis does not pass through the reflective surface. This misalignment can cause vertical fluctuations in the detection point as tilting occurs. Consequently, the difference between z_s and d_f also changes. Moreover, achieving perfect alignment, as discussed in Sect. 7.6.4 can further affect the relationship between z_s and d_f when adjusting θ_tilt and ϕ_tilt.

To solve this problem, we can use Eq. (7.10) to calculate the spot radii of all collected DIs, where the spot radius corresponding to $\mathbf{D}_{i,j,k}$ is denoted as $r_{\mathrm{sp},i,j,k}$. The relationship between $r_{\mathrm{sp},i,j,k}$ and $z_{\mathrm{s},i}$ for two particular combinations of tilting angle and direction are plotted in Fig. 7.21, where the red and blue dots denote the spot radii for two different surface orientations. As illustrated, a v-shape line defined by

$$r_\mathrm{sp} = \begin{cases} r_{\mathrm{sp}0} + m_\mathrm{left}(z_\mathrm{s} - z_{\mathrm{s}0}), & z_\mathrm{s} \le z_{\mathrm{s}0} \\ r_{\mathrm{sp}0} + m_\mathrm{right}(z_\mathrm{s} - z_{\mathrm{s}0}), & z_\mathrm{s} > z_{\mathrm{s}0} \end{cases} \tag{7.20}$$

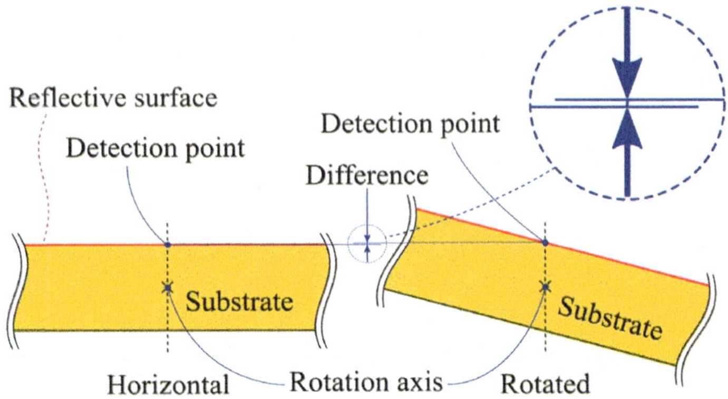

Fig. 7.20 The surface height is tilting-dependent if the rotation axis does not pass through the reflective surface

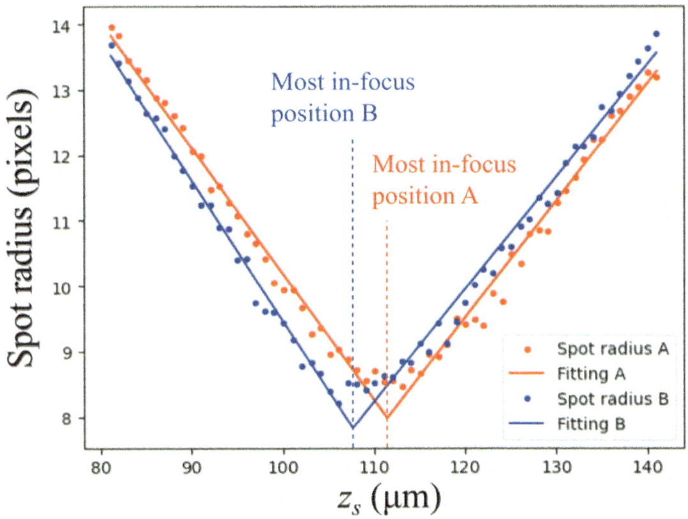

Fig. 7.21 Two example relationships between the spot radius and the feedback values of the z-stage

is used to fit the experimental data. Where z_{s0}, r_{sp0}, m_{left}, m_{right} are the fitting coefficients. (z_{s0}, r_{sp0}) is the coordinate of the vertex. m_{left}, m_{right} are the slopes at the left-hand side and right-hand side of the vertex. The abscissa of the vertex, z_{s0}, is considered to be the most in-focus position (e.g., the most in-focus position A and B in Fig. 7.21), which corresponds to the point of $d_f = 0$, i.e.:

$$d_f = z_s - z_{s0}. \tag{7.21}$$

Be aware that the term z_{s0} also depends on the tilting angle and direction. As shown in Fig. 7.21, data A is collected with $\theta_{tilt} = 0°$. However, when only the BSM is tilted by an angle of 5°, the position of the most in-focus also changes. Hence, for each combination of tilting angle and direction, the corresponding most in-focus position, z_{s0}, must be solved individually, denoted by $z_{s0,j,k}$. Therefore, the final database is a collection of $\mathbf{D}_{i,j,k}$ labeled by $d_{f,i}$, $\theta_{tilt,j}$, and $\phi_{tilt,k}$, where $d_{f,i} = z_{s,i} - z_{s0,j,k}$.

References

1. http://www.korin-hayashi.com.cn/en/content/UV13/438.html
2. http://www2.optics.rochester.edu/workgroups/cml/opt307/spr05/john/
3. Wu G-W, Chen L-C (2023) Resolving measurement ambiguity in diffractive image microscopy for 6DOF surface measurement using designed aberration and multiple-layer perceptron. Int J Adv Manuf Technol. https://doi.org/10.1007/s00170-023-11147-0
4. Wu G-W, Chen L-C (2021) Precise 3-D microscopic profilometry using diffractive image microscopy and artificial neural network in single-exposure manner. Opt Lasers Eng 147:106732
5. Wu G-W, Nguyen DT, Chen L-C (2019) Novel full-field micro surface profilometry using diffractive image correlation without vertical scanning. Opt Lett 44(14):3534–3537
6. Wu G-W (2022) Theory and implementation of microscopic diffractive image profilometry. Ph.D. thesis, National Taiwan University

Chapter 8
Applications of Diffractive Image Profilometry (DIP)

8.1 Introduction

This chapter starts with the basic properties of the actual setup system, progressing from DIs without aberration film to DIs with aberration film. Subsequently, it delves into the testing of system stability. Following this, flat roughness samples are tested to provide a suggested limitation about applicable roughness. Finally, the chapter explores several samples through various methods and conditions in case studies and following discussions.

8.2 DIs Without Aberration Film

Once the system has been set up and well-aligned, the DIs at the center of the FOV are observed first. Subsequently, the DMD is configured as a pinhole with $p_s = 2$, and the BSM is set as horizontal. Further detailed settings are listed in Table 8.1. With these settings, the z-stage facilitates the movement of the entire BSM for vertical scanning to record DIs at different axial positions. Figure 8.1 shows several observed DIs, with the upper row illustrating the intensity in linear scale and the lower row illustrating the intensity in logarithmic scale. Notably, the DIs exhibit certain asymmetric features even without the insertion of the aberration film. The root cause of this inherent aberration is primarily due to the several thin mirrors (1 mm thick) used for 3-D folding. These mirrors lack the flatness and stiffness characteristic of common optical grade mirrors with thick substrates, thus inducing some aberration. To assess the extent of deviation of the setup system from the diffraction limit, the DI at the most in-focus position is fitted by the two-dimensional Gaussian surface expressed as Eq. (6.10), yielding an averaged standard deviation of 5.09 μm, corresponding to a FWHM of 11.99 μm. In comparison, the estimation according to Eqs. (7.5)–(7.7)

L. Chen et al., *Diffractive Image Microscopy for 3D Imaging*, Springer Tracts in Mechanical Engineering, https://doi.org/10.1007/978-981-97-7782-2_8

Table 8.1 Detailed settings for observing the DIs at the center of the FOV

Items	Values	Units
Pinhole size, p_s	2	Micromirror
Source width (square shape)	15.12	μm
Xenon lamp power	100	%
Exposure time	75	μs
Minimum d_f	– 51.05	μm
Maximum d_f	48.87	μm
Nominal d_f sampling pitch	1	μm
d_f scanning steps	101	–
DI average times	10	–
Aberration film	NA	–

is 11.55 μm ($\eta_{safe} = 1$), which is 96.33% of the observed value. Consequently, the setup system can be concluded as close to the diffraction limit (Fig. 8.1).

The next step involves assessing the potential ambiguity problem. However, this problem may not arise due to the inherent aberration. To verify this, the similarity matrix of observed DIs is computed and shown in Fig. 8.2, with four particular cross-sections depicted in Fig. . It can be found that while double peaks can exist like Fig. (a), the heights of the two peaks are easily distinguishable. Instances of two peaks with the same height, as shown in Fig. 5.5, are absent. However, when the defocus value is small, as depicted in Fig. (d), the peak exhibits a "side swell". This phenomenon corresponds to the "x-shape" plateau within the square region in Fig., potentially leading to instability in determining the peak position during real measurements with extra noise. Therefore, despite the absence of expected ambiguity, understanding the influence of an aberration film on the DIs and similarity matrix remains intriguing (Figs. 8.2 and 8.3).

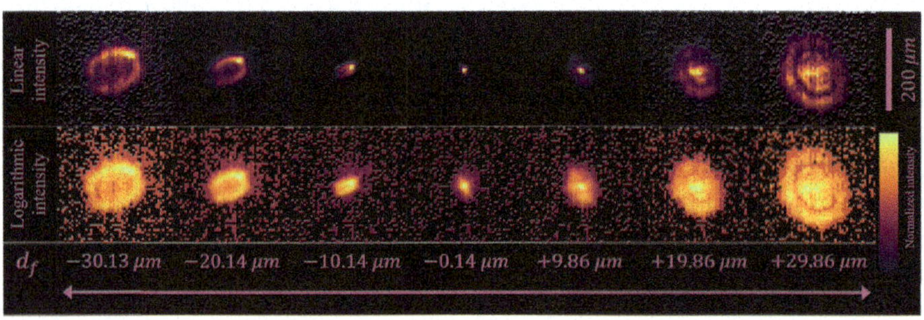

Fig. 8.1 Selected DIs at the center of the FOV. The upper row illustrates the intensity on a linear scale. The lower row illustrates the intensity in the logarithmic scale

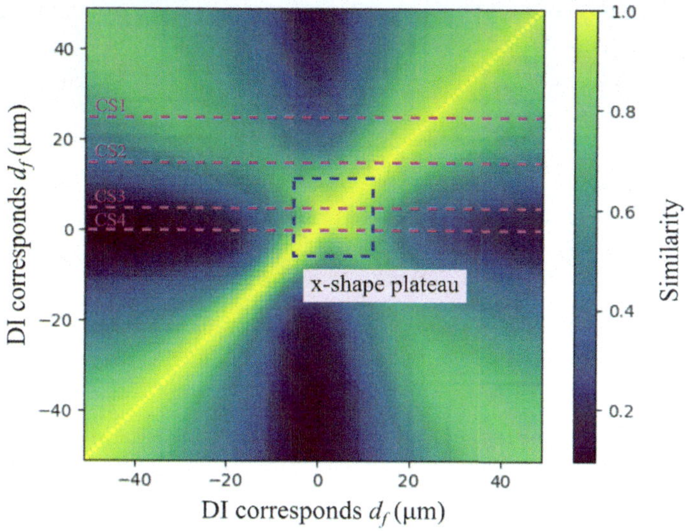

Fig. 8.2 A matrix that reveals the similarities between arbitrary two observed DIs. Four particular cross-sections CS1, CS2, CS3, and CS4 are depicted in Fig.

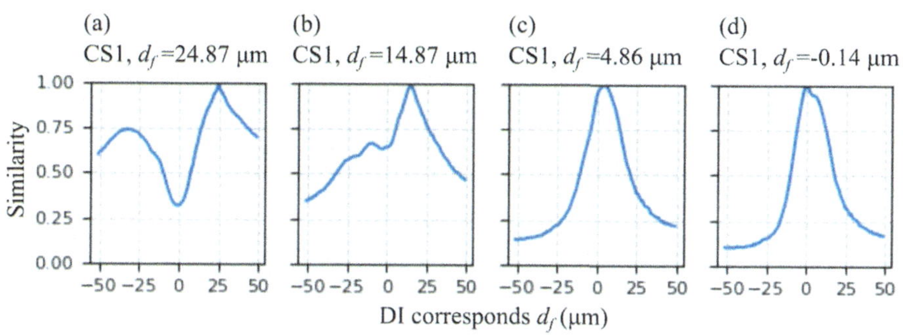

Fig. 8.3 Four cross-sections that correspond to CS1, CS2, CS3, and CS4 in Fig. 8.2

8.3 DIs with Aberration Film

To examine the impact of an aberration film on the DIs, an aberration film with $l_{\text{film}} = 63.5$ mm and $w_{\text{film}} = 60$ mm is inserted, and the same experiment is conducted [1]. Figure 8.4 shows several selected DIs, and the similarity matrix is shown in Fig. 8.5. As observed, the extra aberration induces changes in the DIs. Also, the fitted standard deviation of the most in-focus DI increases to 6.35 μm, corresponding to a FWHM of 14.96 μm. Referring to Fig. (d), the "side swell" beside the peak is much lower for small defocus values, thereby minimizing its potential impact on peak determination. Therefore, peak determination is expected to be more robust in

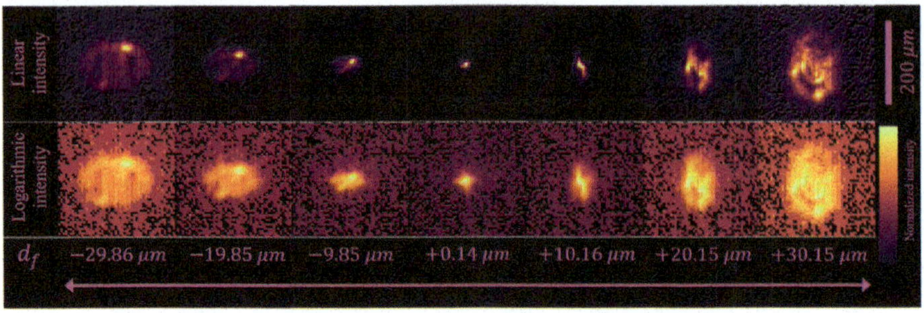

Fig. 8.4 The selected DIs with aberration film. The upper row illustrates the intensity in the linear scale. The lower row illustrates the intensity in the logarithmic scale

real measurements with the aberration film. At the same time, the increase in the most in-focus spot size is about 25%, which remains acceptable if the aberration film is essential (Fig. 8.6).

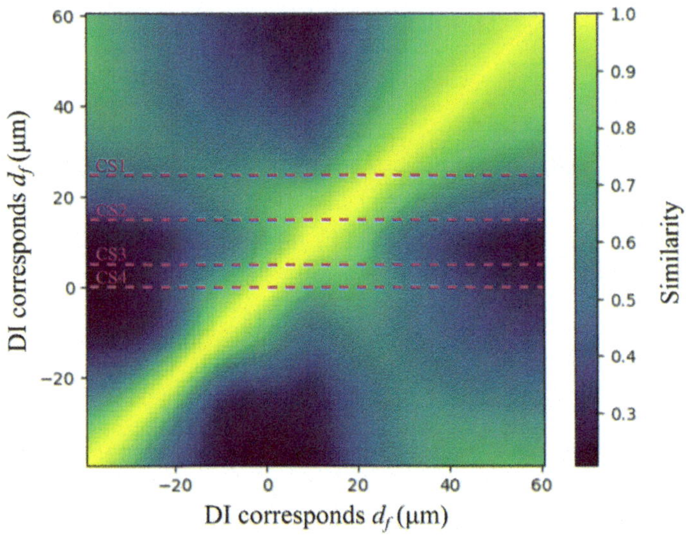

Fig. 8.5 A matrix that reveals the similarities between two arbitrary observed DIs. Four cross-sections CS1, CS2, CS3, and CS4 are depicted in Fig.

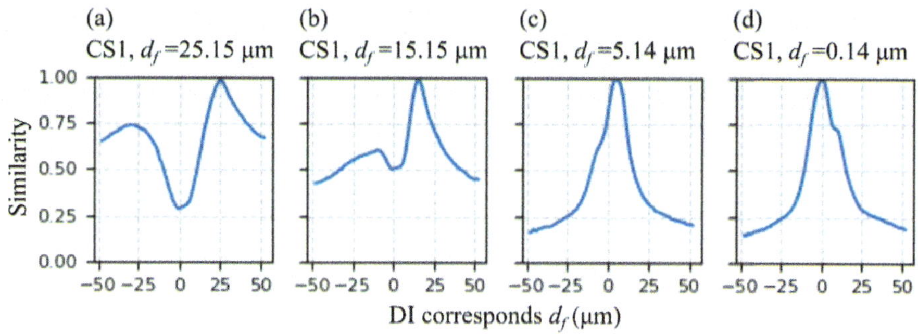

Fig. 8.6 Four cross-sections that correspond to CS1, CS2, CS3, and CS4 in Fig. 8.5

8.4 System Temporal Stability

System stability is essential for any precision measuring instrument. If the setup system drifts over time, the reliability of a database constructed long ago may be compromised. Therefore, knowledge about the system's stability over time is necessary for many purposes. To study this, the system is warmed up for one hour. Subsequently, the DMD projects a single measurement point onto a mirror, and the position of the DI is periodically recorded for 8 h. During this period, the ambient temperature is maintained at 22–23 °C. The result is shown in Fig. 8.7, wherein the DI position is determined using its centroid, and the spot radius is calculated according to Eq. (7.10). As observed, the system exhibits a continuous drift in one direction. However, the drifting of the DI position over this 8-h period is merely around one-fifth of the pixel size (5.5 μm), making it negligible. Furthermore, the change in spot size is about 4.5 μm, translating into an axial variation on the object side equal to a defocus value of 2.14 μm. The primary cause of this drift is inferred to be slow thermal deformation. Fortunately, the defocus drift only causes an axial offset in a measured profile. With the above testing, the thermal sensitivities of the setup system are:

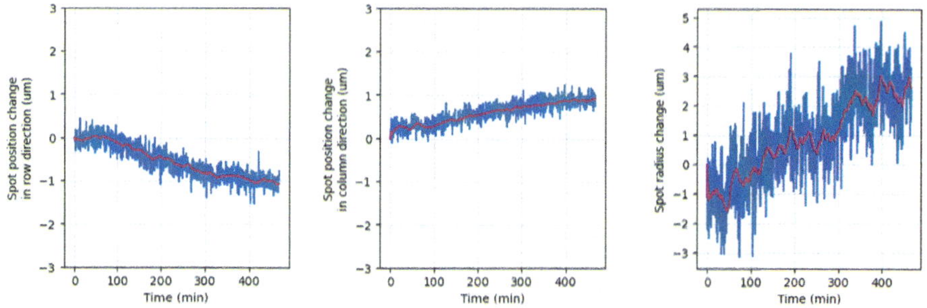

Fig. 8.7 System stability test via monitoring the position and radius of a DI. The magenta line is the 15-min average line

i. DI center change in the row direction is about 0.2 pixels per degree Celsius.
ii. DI center change in the column direction is about 0.2 pixels per degree Celsius.
iii. Change in focusing (axial direction) is about 2.14 μm per degree Celsius.

In conclusion, the setup system is stable enough for a database-based method.

8.5 Measurement Stability

In this section, a flat mirror is measured to acquire the measured height variation at each field point and the uniformity across the whole FOV. To accomplish this, the DMD is configured to conduct a scanning plan of $p_s = 2, p_p = 50$, and $s_p = 50$. This scanning plan comprises a total of 1581 measurement points, with each point having its own database and being resolved individually. Seven databases are built with different sampling pitches to study the influence of sampling pitches on measured height variation. All detailed settings of this experiment can be found in Table 8.2.

First, Fig. 8.8a shows a map illustrating the standard deviation of the measured height (SDMH), with a sampling pitch of 4.0 μm used in the database. It is observed that the distribution of the SDMH is irregular and lacks uniformity across the whole FOV. This irregularity stems from the non-uniform aberration across the FOV,

Table 8.2 Detailed settings for flat mirror measurement

Classification	Items	Values	Units
Scanning plan	p_s	2	Micromirror
	p_p	50	Micromirror
	s_p	50	Micromirror
	Total points in FOV	1581	–
	Total exposure times	1	Frame
Camera settings	Exposure time	67	μs
Database	Type	Reduced database	–
	d_f range	± 30	μm
	d_f sampling pitch	4.0, 2.0, 1.0, 0.5, 0.2, 0.1	μm
Algorithm	Method	IDB with cubic polynomial fitting and interpolation	–
	Interpolation region	5	μm
Others	FOV	1.21 × 1.94	mm
	Objective NA	0.28	–
	Objective DOF	3.5	μm
	Aberration film	NA	–

resulting in different similarity response curves (SRC) at different field points. Consequently, the stability of SRC peak determination varies with field points. From a statistical perspective, Fig. 8.8b shows the histogram and cumulative curve of the SDMH. The mean of SDMH is computed to be 0.138 μm, with 95% of the field points having SDMH below 0.45 μm. Similarly, in Figs. 8.9, 8.10 and 8.11, the SDMH is depicted for databases with sampling pitches of 3.0, 2.0, and 1.0 μm, respectively. Additionally, Fig. 8.12 presents more experimental data, offering further insights into the relationship between the average SDMH and sampling pitch. According to Fig. 8.12, it is found that SDMH decreases rapidly when the sampling pitch approaches the DOF (3.5 μm). When the sampling pitch is below about two-thirds of the DOF, the SDMH remains at approximately 0.03 μm, accounting for only 0.857% of the DOF, resulting in highly stable measured heights.

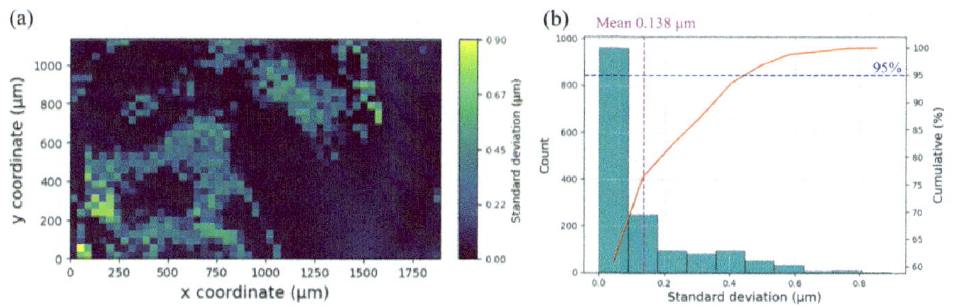

Fig. 8.8 **a** A map of the standard deviation of the measured height and **b** the corresponding histogram and its cumulative curve. The used database has a sampling pitch of 4.0 μm

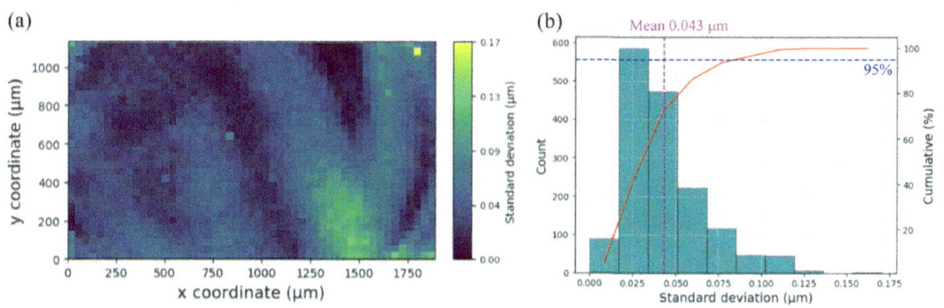

Fig. 8.9 **a** A map of the standard deviation of the measured height and **b** the corresponding histogram and its cumulative curve. The used database has a sampling pitch of 3.0 μm

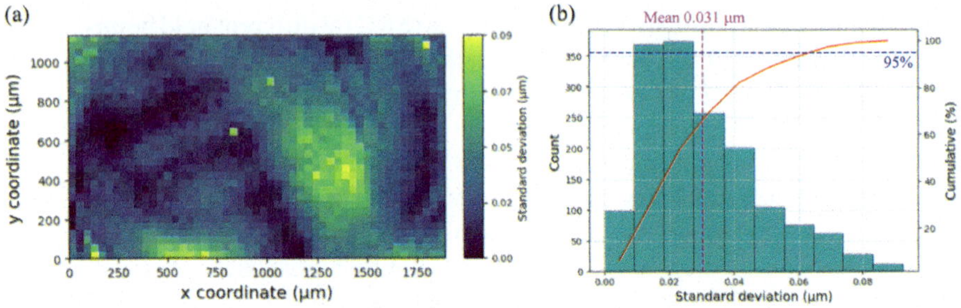

Fig. 8.10 **a** A map of the standard deviation of the measured height and **b** the corresponding histogram and its cumulative curve. The used database has a sampling pitch of 2.0 μm

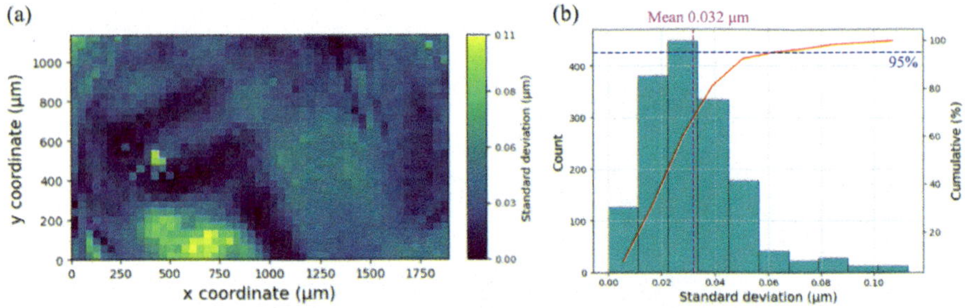

Fig. 8.11 **a** A map of the standard deviation of the measured height and **b** the corresponding histogram and its cumulative curve. The used database has a sampling pitch of 1.0 μm

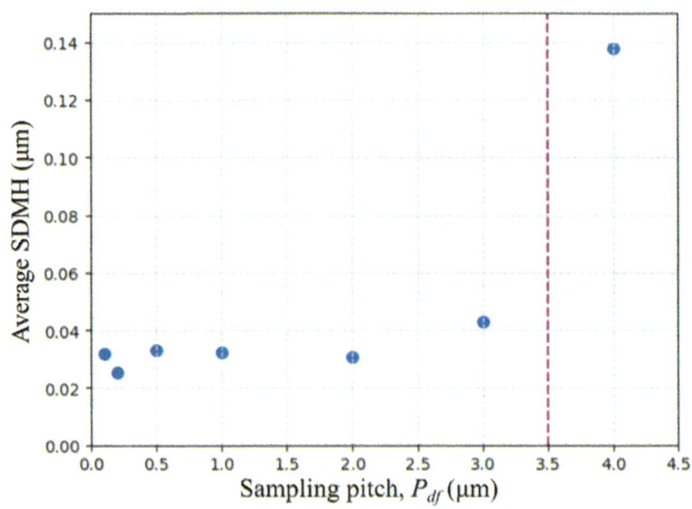

Fig. 8.12 The relationship between the average SDMH and the sampling pitch. The magenta dash line denotes the DOF of the used objective

8.6 Rough Surface

As mentioned in Sect. 5.5, surface roughness can degrade DIs, leading to deteriorated results. To verify this experimentally, DIs were observed for four grinding surfaces with different roughness (Flexbar roughness standards, 16008-CAL), as shown in Fig. 8.13. It is observed that DIs gradually deteriorate with increasing surface roughness. By applying normalized cross-correlation (NCC) to gauge the similarity between all DIs and those corresponding to the specular surface, seven similarity curves can be plotted, as shown in Fig. 8.14. These curves correspond to seven different defocus values. Notably, even with a surface roughness as low as Ra 0.05 μm (a gentle matte surface to the human eye), the similarity can drop below 0.8. However, the similarity value lacks significant interpretive value. To acquire a more practical criterion, flat surfaces with different roughness need to be evaluated. The system settings are listed in Table 8.3. The profiles measured for various roughness levels are shown in Figs. 8.15, 8.16, 8.17, 8.18 and 8.19. Additionally, the relationship between the height standard deviation (outlier removed in advance) and the surface roughness are depicted in Fig. 8.20, where the slope and intercept of the fitted line are 20.252 and 0.296 (μm), respectively. This relationship can aid in estimating whether an object can be measured with an acceptable height error variation. For example, if the standard deviation of height error should be 1 μm or less, then the maximum measurable roughness is about Ra 0.05 μm.

8.7 Step Height

The same system setting as listed in Table 8.2 is used to measure a specular step height sample. Four grooves are measured, as shown in Figs. 8.21 8.22, 8.23 and 8.24. Within these figures, points within the left rectangle box are fitted by a plane, followed by the calculation of the average distance from this plane to the points within the right rectangular box. All result is listed in Table 8.4. The reference depth is the measured result by a commercial laser confocal microscope (KEYENCE VK-X1000). Notably, the largest difference is 0.13 μm when measuring the deepest groove, which is only 0.5% of a 25.3 μm groove. However, as the groove depth decreases, the error ratio increases because the measured difference is independent of the groove depth, thus resulting in an increased error ratio.

8.8 Semiconductor Microbumps

Given the effectiveness of this method with specular samples, a semiconductor microbump chip is chosen for measurement [2]. The microbumps on this chip have a flat-top feature, and the surface finish resembles a high-quality mirror. Because

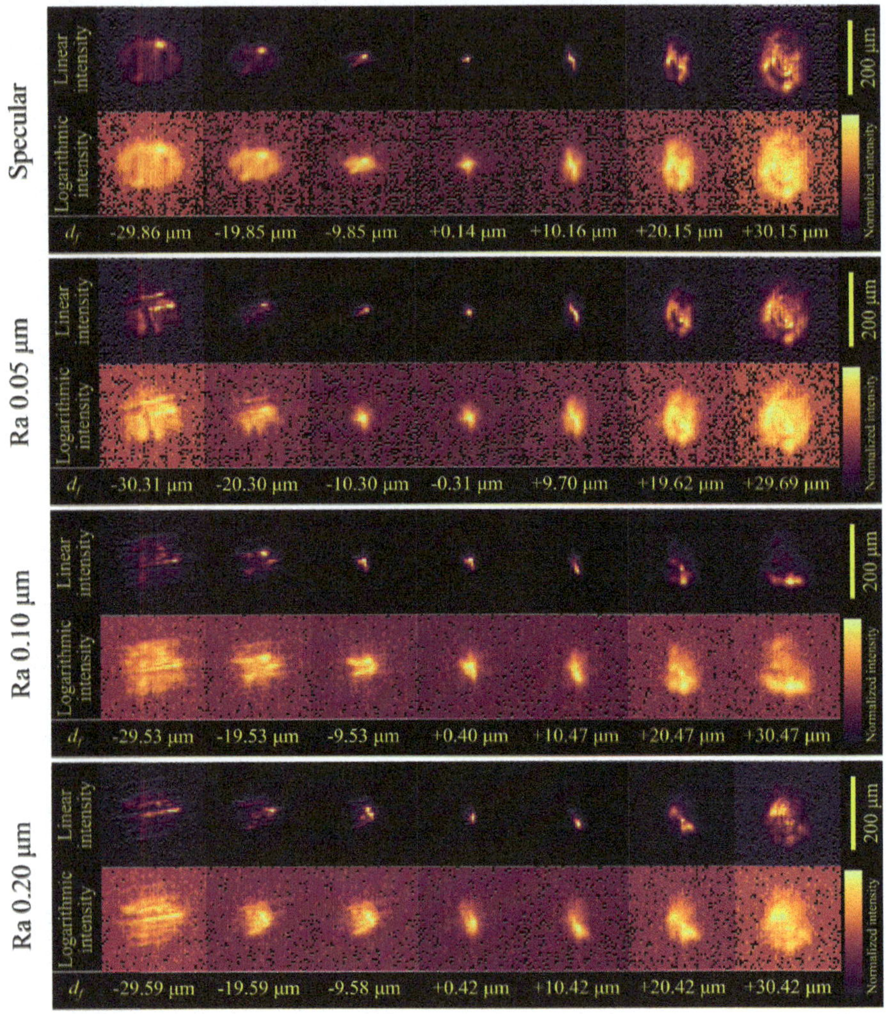

Fig. 8.13 The influence of surface roughness on DIs

of its small size, the objective has been replaced with a 50× objective (Mitutoyo 378-805-3, NA0.42, DOF 1.6 μm), with additional parameters listed in Table 8.5. All preparation procedures are identical to measuring the step height sample. Consequently, Fig. 8.25 shows the measured profile, illustrating a region of 90 × 90 μm with nine microbumps with a nominal pitch of 35 μm and a nominal height of 3 μm, respectively.

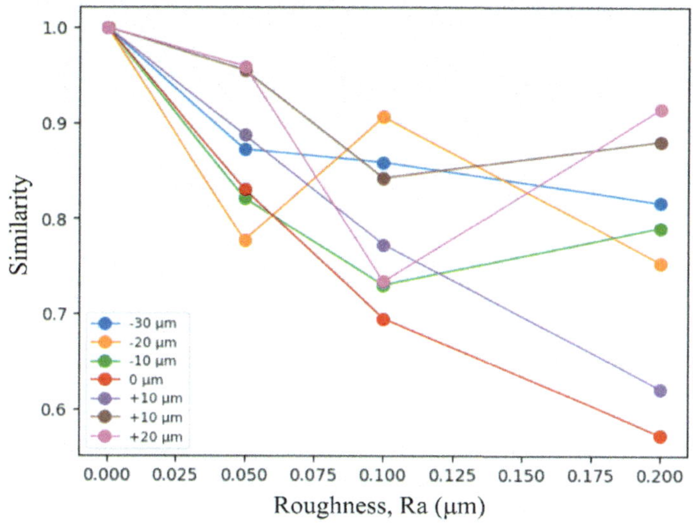

Fig. 8.14 The influence of surface roughness on DIs, measured by the NCC algorithm, compared with the cases corresponding to specular surface

Table 8.3 Detailed settings for flat mirror measurement

Classification	Items	Values	Units
Scanning plan	p_s	2	Micromirror
	p_p	50	Micromirror
	s_p	10	Micromirror
	Total points in FOV	39,525	–
	Total exposure times	25	Frame
Camera settings	Exposure time	50	Ms
Database	Type	Reduced database	–
	d_f range	± 30	μm
	d_f sampling pitch	1	μm
Algorithm	Method	Image comparison	–
	Median filtering (post-processing)	Kernel size 3	Pixel
Others	FOV	1.21 × 1.94	mm
	Aberration film	63.5 bent into 60	mm

8.9 Spherical Mirror

A spherical convex mirror (Edmund #87-664) is chosen as a test object to assess the validity of the proposed methods on surface orientation detection [3]. The convex mirror has a radius of curvature of 400 mm (EFL 200 mm), and a region around

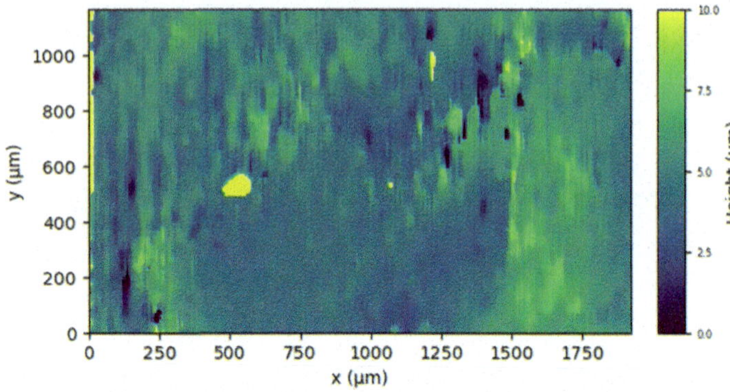

Fig. 8.15 The measured profile of a flat surface with surface roughness of Ra 0.05 μm

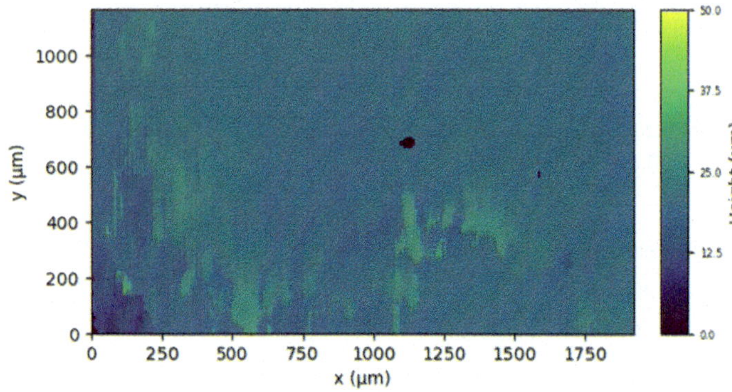

Fig. 8.16 The measured profile of a flat surface with surface roughness of Ra 0.10 μm

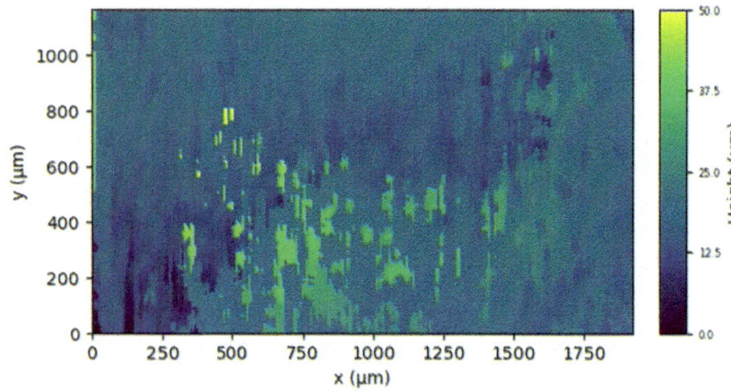

Fig. 8.17 The measured profile of a flat surface with surface roughness of Ra 0.20 μm

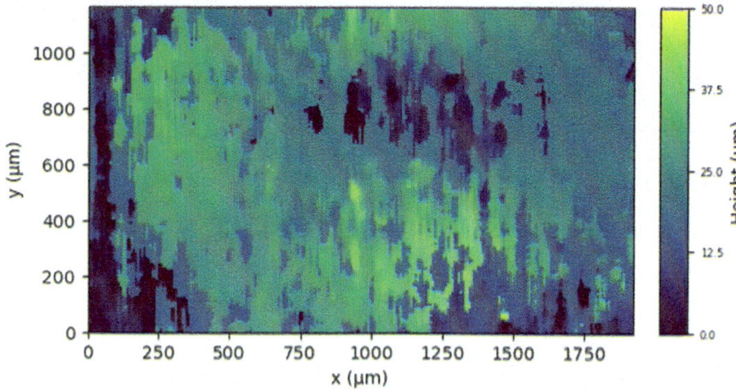

Fig. 8.18 The measured profile of a flat surface with surface roughness of Ra 0.40 μm

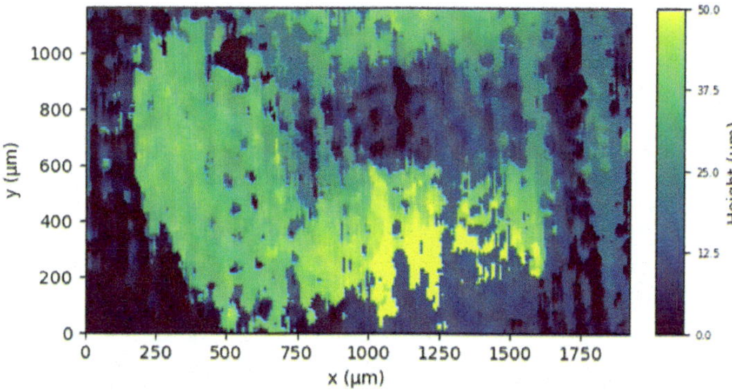

Fig. 8.19 The measured profile of a flat surface with surface roughness of Ra 0.80 μm

its vertex will be measured. Initially, a series of DIs are collected according to the sampling plan described in Table 8.6. Subsequently, these collected DIs are processed into various databases, including the image database (IDB), Gaussian database (GDB), fractional polynomial database (FDB), and PCA database (PDB). Unlike previous experiments utilizing reduced databases, only one pinhole is configured on the DMD to build the database in this scenario. This pinhole is located at the center of the FOV, with a size of 2 μm. Moreover, the lateral scanning is done by the $x - y$ stage, with a 1 mm pitch and measurements taken at 23 × 23 regular grid points, centered at the vertex. Notably, the aberration film is absent during this experiment.

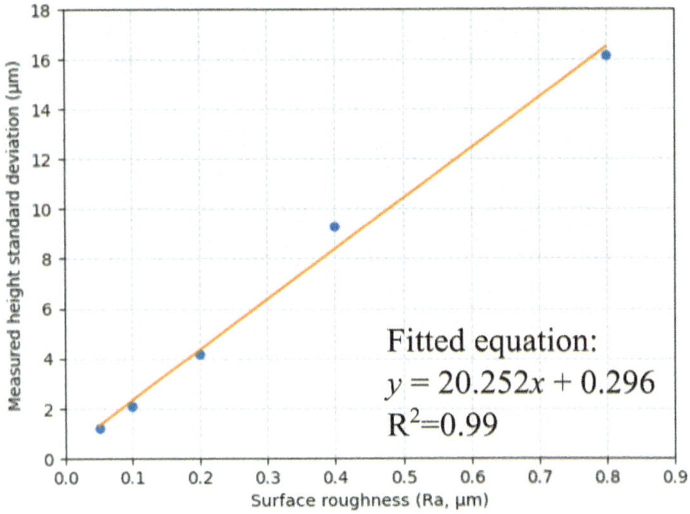

Fig. 8.20 The relationship between the measured height standard deviation and the surface roughness

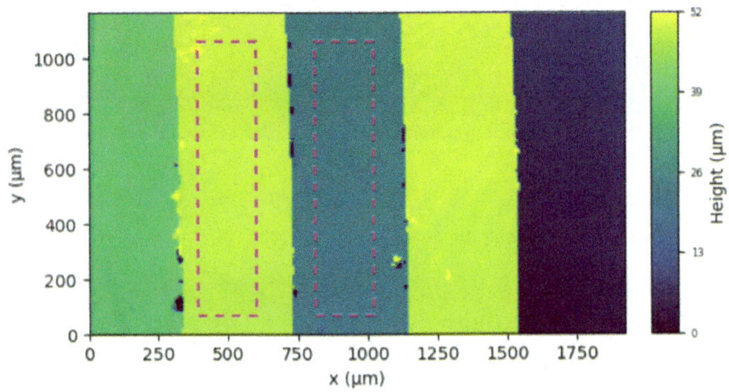

Fig. 8.21 Step height measurement, center at a groove with a 25.30 μm depth

8.9.1 MNCC and MZNCC Method

The measured results using the maximum normalized cross-correlation (MNCC) and maximum zero mean NCC (MZNCC) methods are shown in Fig. 8.26. In the figure, the first row is the measured surface profile, the second row is the measured tilting angle, and the third row is the measured tilting direction. Each of the four columns corresponds to measurements based on different types of databases. From Fig. 8.26, it becomes evident that the result based on IDB has the best performance. This is not surprising because the original DIs are fully preserved. Consequently, results

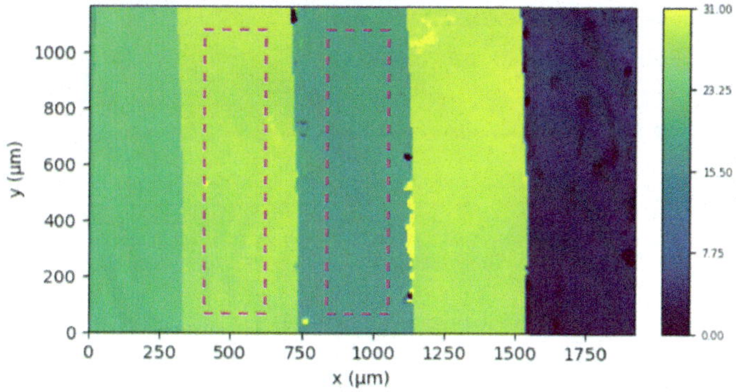

Fig. 8.22 Step height measurement, center at a groove with a 10.20 μm depth

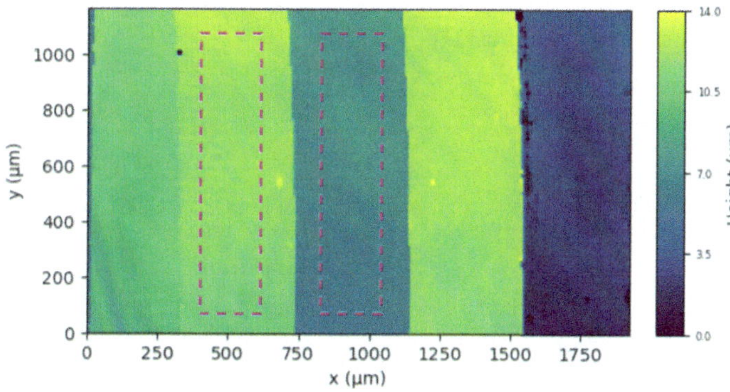

Fig. 8.23 Step height measurement, center at a groove with a 5.10 μm depth

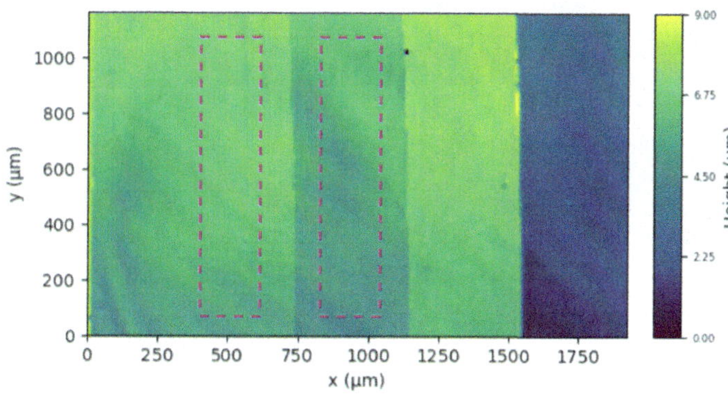

Fig. 8.24 Step height measurement, center at a groove with a 1.20 μm depth

Table 8.4 Measurement result of step height

Groove No.	Reference depth (μm)	Measured average depth (μm)	Standard deviation (μm)	Difference (μm; %)
1	25.30	25.17	0.93	− 0.13; − 0.5%
2	10.20	10.17	0.70	− 0.03; − 0.3%
3	5.10	5.18	0.33	0.08; 1.6%
4	1.20	1.23	0.13	0.03; 2.5%

Table 8.5 Detailed settings for microbumps measurement

Classification	Items	Values	Units
Scanning plan	p_s	2	Micromirror
	p_p	24	Micromirror
	s_p	3	Micromirror
	Total points in FOV	447,744	–
	Total exposure times	64	Frame
Camera settings	Exposure time	5000	μs
Database	Type	Reduced database	–
	d_f range	± 15	μm
	d_f sampling pitch	0.157 (10% of the DOF)	μm
Algorithm	Method	Image comparison	–
	Median filtering (post-processing)	Kernel size 5	Pixel
Others	FOV	0.24 × 0.39	mm
	Aberration film	NA	mm

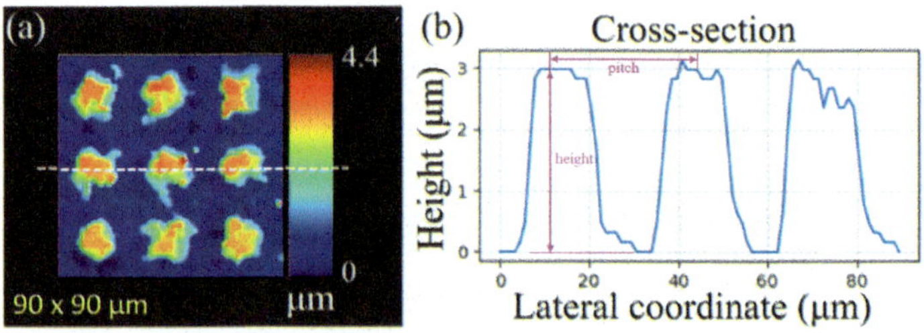

Fig. 8.25 Measured profile of a semiconductor microbump chip

Table 8.6 Sampling plan for the database used to measure a spherical convex mirror

Posture parameter (unit)	Minimum	Maximum	Pitch
Defocus value, d_f (μm)	− 50	50	1
Tilting angle, θ_{tilt} (°)	0	4	1
Tilting direction, ϕ_{tilt} (°)	0	350	10

based on alternative database types cannot provide better results, although they may require less storage space or computational load. However, upon initial observation, it becomes apparent that the FDB and PDB perform inadequately, as the tilting angle and direction appear disorganized. The failure of measurements based on FDB could be attributed to improper sensitivities of the fitted parameters to the ZNCC values. Furthermore, the PDB uses only 5 parameters to represent the original DIs, whereas other methods utilize 7 parameters, potentially leading to information loss and consequently yielding worse results.

Regarding the measured profile, Fig. 8.26a–d provide reasonable results but with different random errors. This can be estimated by fitting the measured profiles with a spherical surface expressed by

$$z(x, y) = \sqrt{R_{\mathrm{curv}}^2 - (x - x_{\mathrm{c}})^2 - (y - y_{\mathrm{c}})^2} + z_{\mathrm{c}}, \tag{8.1}$$

where x and y are the lateral coordinates, $(x_{\mathrm{c}}, y_{\mathrm{c}}, z_{\mathrm{c}})$ is the center coordinate of the fitted sphere, and R_{curv} is the radius of the fitted sphere. The fitting residual root mean square (FRRMS) is then calculated to be 1.023, 3.435, 1.252, and 8.491 μm, respectively. In comparison, the DOF of the used objective is 3.5 μm. Therefore, the ratio of these FRRMS to the DOF is 29.23, 98.14, 35.77, and 242.60%.

With regard to the measured tilting angle, as shown in Fig. 8.26e–h, they are fitted by

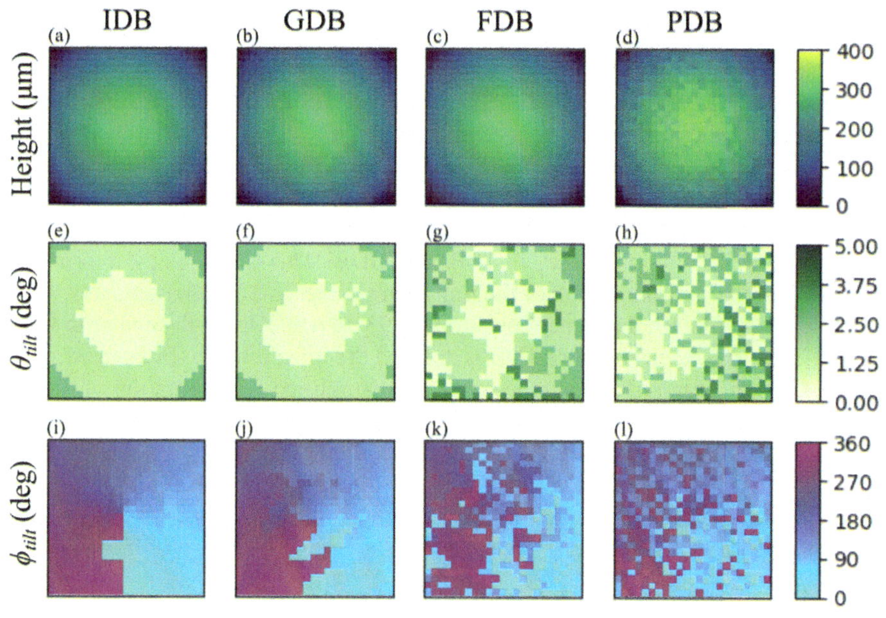

Fig. 8.26 MNCC and MZNCC measurement result of a spherical convex mirror

$$\theta_{\text{tilt}}(x, y) = \sin^{-1}\left(\frac{\sqrt{(x - x_c)^2 + (y - y_c)^2}}{R_{\text{curv}}}\right). \tag{8.2}$$

However, it is found that the fitted center deviates from the one given by the profile fitting. For the case of IDB, the distance between the fitted centers of the profile and the tilting angle is about 719.39 μm, indicating a large difference.

This phenomenon can be explained with a simpler 2-D case. Consider Fig. 8.27, where the y-axis is the optical axis of the DIP system. However, the moving direction of the stage is not perpendicular to the optical axis but follows a direction defined by $y = mx$. Let the initial surface of the detected mirror be $y = \sqrt{R_{\text{curv}}^2 - x^2}$, then the measured profile will be

$$y = \sqrt{R_{\text{curv}}^2 - x^2} - mx. \tag{8.3}$$

The measured "vertex" position can be found by solving

$$\begin{aligned}
&\frac{\mathrm{d}y}{\mathrm{d}x} = 0 \\
&\Rightarrow \frac{-x}{\sqrt{R_{\text{curv}}^2 - x^2}} - m = 0 \\
&\Rightarrow x = -\sqrt{\frac{m^2 R_{\text{curv}}^2}{1 + m^2}} \text{ (choose negative if } m \text{ is positive).}
\end{aligned} \tag{8.4}$$

By substituting $x = 719.39$ μm and $R_{\text{curv}} = 400$ mm into Eq. (8.4), the result m is about 0.01798, corresponding to a climbing angle of 0.103°. This explains that despite careful alignment efforts, a center deviation as large as 719.39 μm persists due to its high sensitivity. However, this can be seen as an unexpected validation of the proposed method, as it demonstrates the method's capability to detect misalignment of this magnitude. Despite the deviated center, the FRRMS of the tilting angle is 0.291° and 0.321° for the cases of IDB and GDB. This discrepancy is primarily attributed to the sampling pitch of the tinting angle, set as 1° in this scenario.

In the case of GDB, the center deviation is 393.56 μm. Although smaller in value, theoretically, the result based on IDB is deemed the most reliable. In contrast, GDB is based on DI fitting parameters, serving as a simplified representation of the raw data. Thus, the smaller value does not hold significant implications. Regarding the

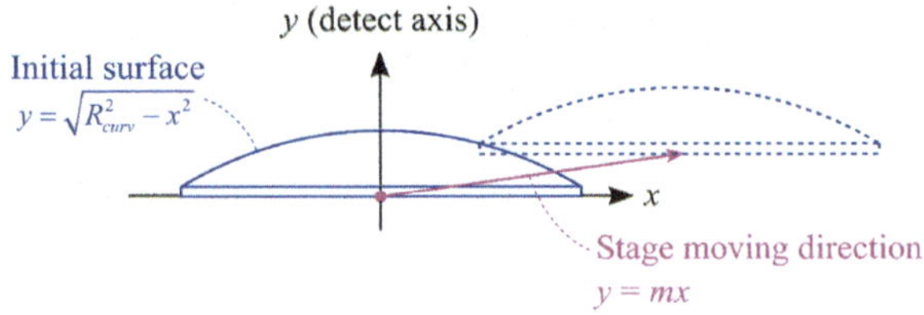

Fig. 8.27 Coordinate definition for center deviation explanation

cases of FDB and PDB, their outcomes are considerably disorganized, making further discussion irrelevant.

Next, about the measured tilting direction, the singular point at the vertex is evident in Fig. 8.26 (i) and (j). This singularity occurs due to the zero slope at the vertex (horizontal), making the tilting direction essentially meaningless and resulting in a discontinuous jump. Also, the boundary around 0° is not as stable as expected, primarily because the values at 1° and 359° are quite different despite their close proximity. Therefore, the tilting direction cannot be directed fitted. Instead, it has to be transformed by a sinusoidal function, i.e.:

$$\gamma_m(x, y) = \sin(\phi_m(x, y)), \tag{8.5}$$

where $\phi_m(x, y)$ is the measured tilting direction and $\gamma_m(x, y)$ is a transformed function. The transformed function is fitted by

$$\begin{cases} \gamma_{tilt}(x, y) = \sin\left(\tan^{-1}(y'/x')\right) \\ \begin{bmatrix} x' \\ y' \end{bmatrix} = \begin{bmatrix} \cos\phi_0 & -\sin\phi_0 \\ \sin\phi_0 & \cos\phi_0 \end{bmatrix} \begin{bmatrix} x - x_c \\ y - y_c \end{bmatrix} \end{cases} \tag{8.6}$$

with ϕ_0 being the reference azimuthal direction. The FRRMS of the tilting direction can be estimated by

$$\phi_{\text{FRRMS}} \cong \sin^{-1}(\gamma_{\text{FRRMS}}), \tag{8.7}$$

where ϕ_{FRRMS} is the FRRMS of tilting direction and γ_{FRRMS} is the FRRMS of the transformed function, given by

$$\gamma_{\text{FRRMS}} = \sqrt{\sum_{x,y} (\gamma_{\text{tilt}} - \gamma_m)^2}. \tag{8.8}$$

With Eqs. (8.5)–(8.8), the FRRMS of tilting direction is estimated to be 6.94° and 14.10° for the IDB and GDB cases. This discrepancy is primarily caused by the sampling pitch of the tilting direction, which is 10° in this scenario.

Finally, the fitted radii of curvature for the four cases are 399.582, 403.481, 399.159, and 401.964 mm, respectively. For comparison, the nominal value is 400 mm, and the CMM (Mitutoyo STRATO-Apex 574) gives a radius of curvature of 397.042 μm. However, as a PZT module is used to keep the mirror surface within the measurable range while the $x - y$ stage moves the mirror laterally, the accuracy of the fitted radii primarily depends on the accuracy of the PZT module and the $x - y$ stage, rather than the proposed method. Conversely, the FRRMS indicates the level of stability achievable by the proposed method, making the FRRMS discussions above more meaningful.

In summary, the above experiment results have been organized into Table 8.7. It can be concluded that the MNCC method, based on IDB, has the best performance.

Table 8.7 A summary table of the MNCC and MZNCC experiments, numbers in brackets are of less meaning

Items (unit)	IDB	GDB	FDB	PDB
Height FRRMS (μm)	1.023	3.435	1.252	8.491
Height FRRM/DOF (%)	29.23	98.14	35.77	242.60
Tilting angle FRRMS (°)	0.291	0.321	0.841	0.932
Tilting direction FRRMS (°)	6.940	14.097	(38.311)	(44.669)
Fitted radius of curvature (mm)	399.582	403.481	399.159	401.964

Furthermore, the MZNCC method, based on GDB, is the only acceptable method that does not use the original DI.

8.9.2 INCC and IZNCC Method

As introduced in Sect. 6.5.1, further interpolation can be applied to acquire smooth and continuous measurement results from a discrete database. Through interpolation and finding the extremes of the fitted hypersurface, the measured results are shown in Fig. 8.28. Unfortunately, as evident from the figure, the utilization of complex equations such as Eqs. (6.5)–(6.7) for fitting the hypersurface increases the likelihood of fitting failures, as indicated by the black points in Fig. 8.28. Therefore, the achieved results are even worse than simply searching within the discrete sampling points for the highest NCC or ZNCC value (i.e., as in the MNCC and MZNCC methods). Based on these findings, it may be inferred that manually designing a proper equation and determining a set of effective initial parameters for fitting problems beyond a certain level of complexity are notably challenging. Therefore, machine learning approaches present an alternative avenue worth exploring if the objective is to obtain a continuous measurement result.

8.9.3 MLP Method

As introduced in Sect. 6.5.3, multiple layer perceptron (MLP) is designed to learn the underlying mapping relationship between the normalized feature vectors (NFVs) and the posture parameters. The MLP is designed to have one input layer, three hidden layers, and one output layer. The number of neurons in these layers is 7, 200, 200, 200, and 3, respectively. Each neuron uses the exponential linear unit (ELU) as the activation function to ensure a smooth mapping relationship. Moreover, all weights and biases are randomly initialized using Xavier's method [4].

Subsequently, the GDB is randomly shuffled and divided into three parts: a training set (72%), a validation set (18%), and a test set (10%). Where the training set is used

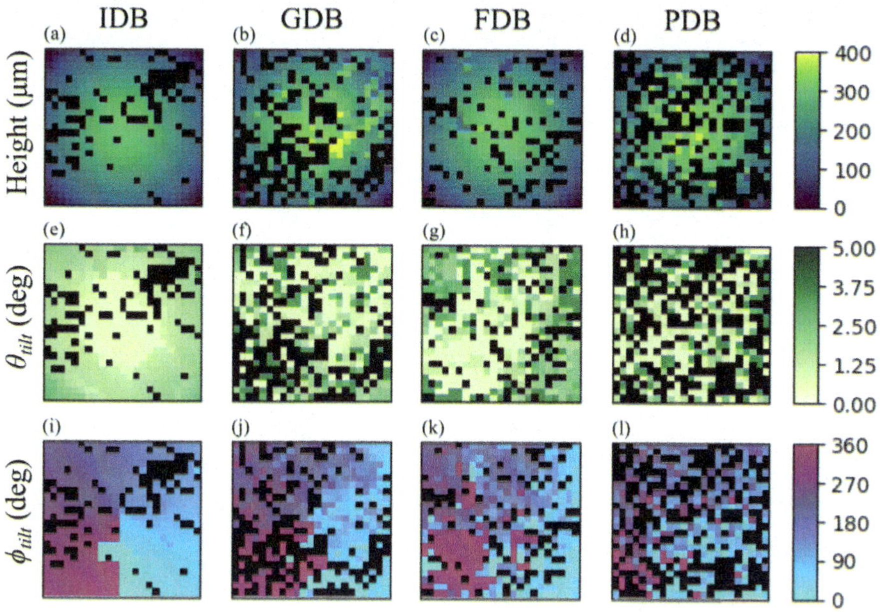

Fig. 8.28 MNCC and MZNCC measurement result of a spherical convex mirror, fitting fail at black points

to train an MLP, i.e., update its weights and biases. The validation set is used to monitor the training process, while the test set evaluates the final performance after the MLP has completed training. These three parts contain 21,708 different NFVs in total.

The training set is further divided into 10 equal size batches to facilitate mini-batch training. These batches are then fed into the MLP one by one. As each batch progresses through the MLP, a series of predicted NPPs are acquired. Subsequently, the partial derivatives of the loss value with respect to all weights and biases are calculated via the backpropagation algorithm. Following this, all the weights and biases are updated according to the Adam [5] method. Upon completion of the tasks by all 10 batches, it constitutes an epoch. The MLP is continuously trained for 100,000 epochs, during which the loss value steadily decreases, as shown in Fig. 8.29.

Once the MLP is trained, the observed DIs obtained during the measurement of the convex mirror are fitted with a two-dimensional Gaussian surface to acquire their NFVs. These NFVs are then fed into the trained MLP, and the resulting output values must be restored back to their original scale according to Eq. (4.5), as the data used to train the MLP is featured normalized. Finally, the restored output gives the measured result, as shown in Fig. 8.30. First, it can be found that the MLP effectively provides interpolated measurement results without invalid points, contrasting Fig. 8.28. This indicates that the MLP has learned a complex function that exceeds human imagination and can smoothly describe the underlying relationship. Second, the FRRMS

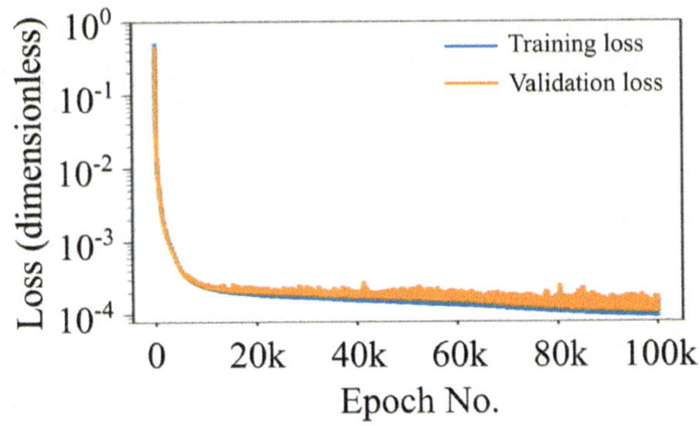

Fig. 8.29 The loss value during the training process

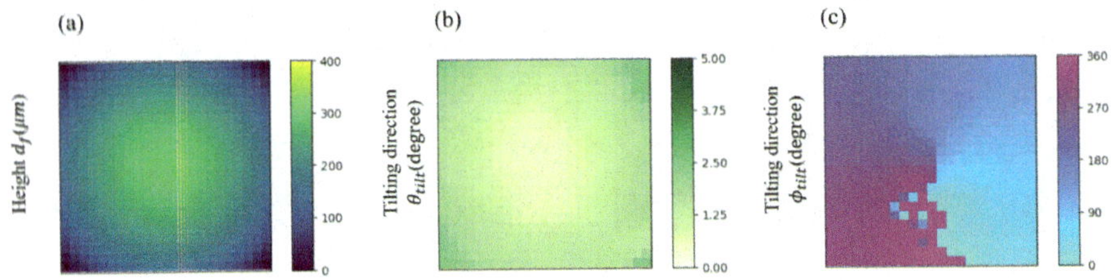

Fig. 8.30 The measured result of a trained MLP

for the profile is 0.903 μm (25.80% of the DOF), and for the tilting angle is 0.151°. Compared to the result of IDB using MNCC, the FRRMS for the profile was reduced by 11.73%, and for the tilting angle was reduced by 48.11%. Thus, these values are even better than the result of MNCC based on IDB. However, the FRRMS of the tilting direction increases to 9.123°, primarily due to instability around the boundary of 0° and 360°, similar to the trend observed in Fig. 8.26 (i) and (j).

Another interesting question arises: can an original DI be inputted into a MLP and be transformed into posture parameters? Theoretically, this is feasible, yet it has not been achieved successfully thus far. The difficulty is attributed to the dilution of meaningful information. When using original DIs, a large portion of the background is preserved instead of solely focusing on critical extracted features. However, despite these challenges, this remains a worthwhile direction to pursue.

8.9.4 Measurement Speed

Although measurement speed is not a primary concern in this book, the use of the DMD enables considerable speed enhancements when measuring samples with a reduced database. For example, in the step height test, the setup system can reach a speed of 2.28 (mm^2/s), whereas the Keyence confocal microscope operates at a speed of 0.286 (mm^2/s) [6]. Therefore, the setup system is approximately 8 times faster than the Keyence confocal microscope. This is attributed to the DMD's ability to open thousands of pinholes at a time, while the Keyence confocal microscope performs point-wise scanning. However, it's noteworthy that the speed increase is not proportional to the number of pinholes opened by DMD because the processing of DIs imposes a much heavier computational load. Therefore, employing skilled programming techniques and hardware acceleration, such as using single instruction multiple data (SIMD), is necessary. Otherwise, the speed may even be slower than that of a point-wise scanning confocal microscope.

References

1. Wu G-W, Chen L-C (2023) Resolving measurement ambiguity in diffractive image microscopy for 6DOF surface measurement using designed aberration and multiple-layer perceptron. Int J Adv Manuf Technol. https://doi.org/10.1007/s00170-023-11147-0
2. Wu G-W, Nguyen DT, Chen L-C (2019) Novel full-field micro surface profilometry using diffractive image correlation without vertical scanning. Opt Lett 44(14):3534–3537
3. Wu G-W, Chen L-C (2021) Precise 3-D microscopic profilometry using diffractive image microscopy and artificial neural network in single-exposure manner. Opt Lasers Eng 147:106732
4. Glorot X, Bengio Y (2010) Understanding the difficulty of training deep feedforward neural networks. In: Proceedings of the thirteenth international conference on artificial intelligence and statistics. JMLR workshop and conference proceedings, pp 249–256
5. Kingma DP Jimmy B (2014) Adam: A method for stochastic optimization. arXiv preprint arXiv: 1412.6980
6. Wu G-W (2022) Theory and implementation of microscopic diffractive image profilometry. Ph.D. thesis, National Taiwan University

Chapter 9
Chromatic Confocal Diffractive Image Profilometry (CC-DIP)

9.1 Introduction

This chapter introduces the core principles of a chromatic confocal diffractive image profilometry (CC-DIP) system, which can be regarded as a variant of DIP presented in previous chapters. CC-DIP aims to enhance the lateral and axial resolution of conventional point scanning chromatic confocal microscopy (PSCCM) with a deep learning-assisted deconvolution algorithm and a scalar diffraction-based parametric model. Similar to DIP, diffractive images from a chromatic confocal microscopy system are captured and processed with a model-based database for practical measurements.

9.2 Derivation of CCM Model

This section proposes a wave theory model for chromatic confocal microscopy (CCM). CCM is characterized by its color dispersion and the selection of a specific wavelength for each position along the axial direction, detected on the camera as a spectrum signal.

When modeling the spectral signal, several design parameters of the CCM system contribute to its performance and functionality. These parameters include various dimensional characteristics, including the sizes of the illumination and detection pinholes, numerical apertures, and the chromatic objective's properties. Understanding the interplay between the spectral signal and the design parameters facilitates advancements in optimizing CCM's optical performance.

A model is developed to describe the monochromatically illuminated axial confocal signal, considering the relationship between displacement and wavelength [1]. This model provides insights into how the confocal signal changes with variations in the axial position within the object plane and adjustments in the illumination wavelength. For an extended object in CCM, the light emitted or reflected from each

L. Chen et al., *Diffractive Image Microscopy for 3D Imaging*, Springer Tracts in Mechanical Engineering, https://doi.org/10.1007/978-981-97-7782-2_9

point on the object can be treated as originating from multiple point sources. These point sources emit rays in various directions. In geometric optics, the rays from a specific point on the object that pass through an ideal imaging system converge to form an image at a corresponding point.

In CCM, the optical layout typically involves multiple lenses, which can be modeled using an "enclosed" imaging model. In this theoretical framework, the propagation of light from the entrance pupil to the exit pupil is primarily analyzed based on the principles of geometric optics. However, it is crucial to acknowledge and incorporate the impact of diffraction phenomena as the light continues its path from the exit pupil toward the image plane. Diffraction effects play a significant role in determining the behavior and characteristics of light, especially when interacting with small-scale structures. Therefore, to develop a comprehensive understanding of the optical system, both geometric optics and diffraction effects must be considered in the analysis.

In scenarios involving coherent imaging, the resulting image is generated by convolving the predicted image with the PSF of the optical system. The PSF describes how a point source in the object space is spread out or blurred when projected onto the image space. This convolution operation accounts for the interaction of the predicted image and the optical system's PSF, determining the final image appearance.

Conversely, when incoherent illumination is employed, the image intensity is obtained by convolving the PSF with the ideal image intensity. The intensity PSF captures the spreading of the illumination source, taking its characteristics into account. By considering these optical principles and utilizing appropriate modeling techniques, it becomes possible to understand the behavior of light in chromatic confocal microscopy systems and accurately interpret the resulting images.

In the specific context of this book, an incoherent light source, such as a Xenon lamp, is used. This choice results in incoherent illumination patterns at different object points, as there is no correlation between the phasor amplitudes. Consequently, the imaging process in CCM becomes partly incoherent.

Figure 9.1 illustrates the schematic of a CCM system, which comprises several key components, including a light source, achromatic lenses (AC), a beam-splitter, a chromatic confocal objective, an illumination pinhole, a detection pinhole, a diffraction grating, and camera.

The illumination source employed is assumed to have a uniform spatial radiance, meaning that the intensity is constant across the entire illumination field. The illumination field can be represented as:

$$I_{\text{ill}} = \begin{cases} 1, \sqrt{x_{\text{ill}}^2 + y_{\text{ill}}^2} \leq r_{\text{ill}} \\ 0, \sqrt{x_{\text{ill}}^2 + y_{\text{ill}}^2} > r_{\text{ill}} \end{cases} \tag{9.1}$$

where r_{ill} represents the radius of the illumination pinhole, which is 15 μm, and $(x_{\text{ill}}, y_{\text{ill}})$ represents the illumination coordinates within the illumination space. The coordinates $(x_{\text{ill}}, y_{\text{ill}})$ correspond to different locations on the object field where the illumination is applied.

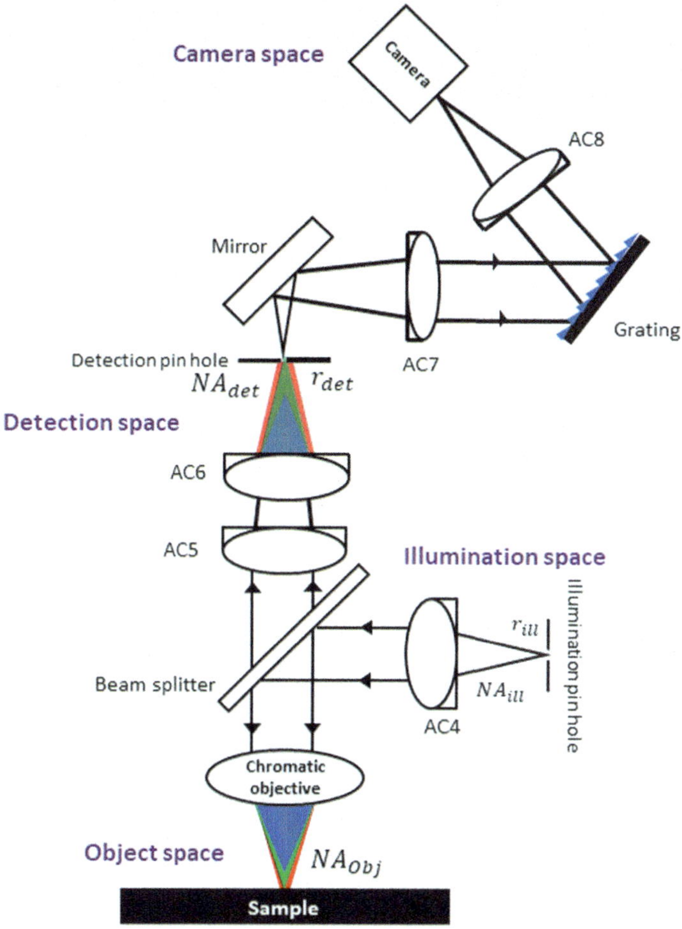

Fig. 9.1 Multispectral focusing principle of chromatic confocal microscopy

Given the Abbe sine condition in the optical system, the lateral magnification from the illumination space to the object space, β_1, can be expressed as:

$$\beta_1 = \frac{\mathrm{NA_{ill}}}{\mathrm{NA_{obj}}} \tag{9.2}$$

where $\mathrm{NA_{ill}}$ and $\mathrm{NA_{obj}}$ is the NA of the illumination and object space, respectively. Similarly, the transverse magnification from the object to the detection space, β_2, can be expressed as:

$$\beta_2 = \frac{\mathrm{NA_{obj}}}{\mathrm{NA_{det}}} \tag{9.3}$$

Table 9.1 Magnification parameter for the entire confocal system

Magnification parameter name	Magnification value
β_1	0.1289
β_2	13.1648
β_3	0.625

where $\mathrm{NA_{det}}$ is the detection space's *NA*. The overall magnification of the illumination-object-detection space is the product of β_1 and β_2, shown as:

$$\beta = \beta_1\beta_2 = \frac{\mathrm{NA_{ill}}}{\mathrm{NA_{det}}} \tag{9.4}$$

Similarly, the transverse magnification from the detection space to the camera space, β_3, can be expressed as:

$$\beta_3 = \frac{\mathrm{NA_{det}}}{\mathrm{NA_{spectrometer}}} \tag{9.5}$$

The lateral magnification from the detection space to the camera space is affected by the numerical aperture ($\mathrm{NA_{spectrometer}}$) of the camera space. Furthermore, the magnification of an optical system varies for different wavelengths. This variation occurs because the focusing distances in the chromatic objective, located in the object space, vary depending on the wavelength from the exit pupil. Table 9.1 presents the magnification of the entire optical system specifically for a wavelength of 500.53 nm.

Using the aforementioned magnification expression, it is possible to determine the spot sizes on the detection and object spaces based on the camera's pixel size. The camera employed in the optical system has a pixel size of 1.85 μm. Further details regarding the camera will be provided in the next chapter.

The numerical aperture (NA) in CCM is wavelength dependent. Consequently, the intensity pattern of the ideal image in the detection space, as predicted by geometrical optics and accounting for the extended illumination source, can be described by the following equation:

$$I_{\mathrm{detector}}(x, y) = \begin{cases} 1, & \sqrt{x^2+y^2} \le \beta r_{\mathrm{ill}} \\ 0, & \sqrt{x^2+y^2} > \beta r_{\mathrm{ill}} \end{cases} \tag{9.6}$$

For a circular pinhole in detection space, the intensity sensitivity can be expressed as a function of the detection coordinates (x, y) with the following equation:

$$D_S(x, y) = \begin{cases} 1, & \sqrt{x^2+y^2} \le r_{\mathrm{det}} \\ 0, & \sqrt{x^2+y^2} > r_{\mathrm{det}} \end{cases} \tag{9.7}$$

where r_{det} is the radius of the detection pinhole.

Figure 9.2 illustrates the schematic of the illumination and the object space, depicting the transition of light rays from the illumination to the object space. The parameters for the illumination-object transition can be found in Table 9.2.

The circular pupil function for the illumination-object space of a diffraction-limited optical system is given by [2]:

$$P_{\text{ill}}(x, y) = \text{circ}\left(\frac{\sqrt{x^2 + y^2}}{R_{XP\text{ill}}}\right) \tag{9.8}$$

where $R_{XP\text{ill}}$ represents half of the exit pupil diameter of the illumination-object space. The coherent transfer function can be expressed as follows [2]:

$$H_{\text{ill}}(f_U, f_V) = \text{circ}\left(\frac{\sqrt{(-\lambda Z_{XP\text{ill}} f_U)^2 + (-\lambda Z_{XP\text{ill}} f_V)^2}}{R_{XP\text{ill}}}\right) \tag{9.9}$$

Or

$$H_{\text{ill}}(f_U, f_V) = \text{circ}\left(\frac{\sqrt{(f_U^2 + f_V^2)}}{f_0}\right) \tag{9.10}$$

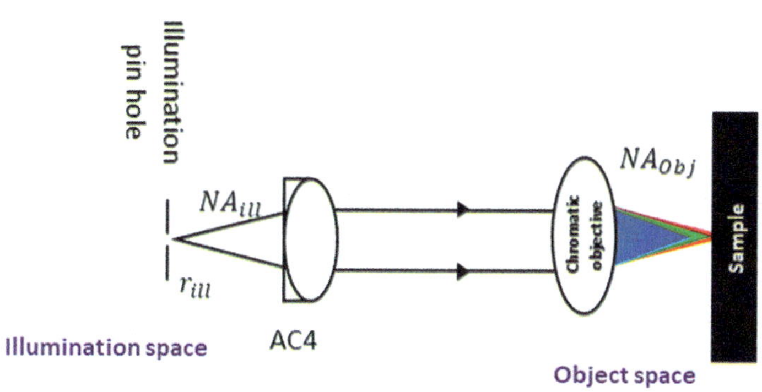

Fig. 9.2 Schematic of the illumination-object space

Table 9.2 Illumination-object space parameters

Parameter	Values	Units
Wavelength (λ)	500.53	nm
Exit pupil diameter ($D_{XP\text{ill}}$)	19,207.49	mm
Exit pupil distance ($Z_{XP\text{ill}}$)	11,139.00	mm
Object space width (square shape)	28.8	μm

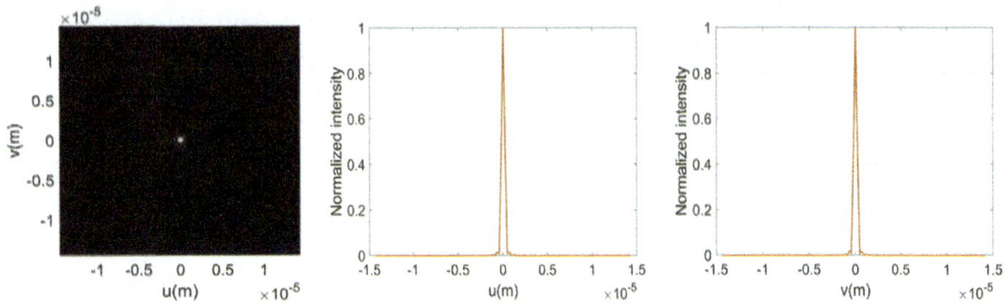

Fig. 9.3 Diffraction-limited illumination PSF and its cross section in object space

where f_0 represents the coherent cutoff frequency, while f_U and f_V denote the frequency coordinates. f_0 is given as:

$$f_0 = \frac{R_{XP\mathrm{ill}}}{\lambda Z_{XP\mathrm{ill}}} \tag{9.11}$$

The PSF or incoherent impulse response, represented by $|h|^2$, can be calculated using the following equation [2]:

$$\mathrm{PSF_{ill}} = |h_{\mathrm{ill}}(u, v)|^2 = \left|\mathcal{F}^{-1}\{H_{\mathrm{ill}}(f_U, f_V)\}\right|^2 \tag{9.12}$$

Figure 9.3 illustrates the obtained PSF in the object space, which is diffraction limited. The PSF represents the optical system's response to a point source and describes the spatial distribution of light in the object space.

To validate the PSF obtained from the wave theory-based parametric model, it is compared with the PSF obtained from Zemax software, as depicted in Fig. 9.4. By comparing the results from the model with those obtained from Zemax, the accuracy and reliability of the model can be evaluated. This comparative analysis serves as a validation step to ensure the alignment between the theoretical predictions and simulations conducted using Zemax.

To perform the convolution between the PSF in the object space and the 30 μm diameter pinhole located in the illumination space, it is essential to transfer the pinhole from the illumination space coordinate system to the object space coordinate system. This transfer is required because the PSF is defined in the object space, and the pinhole's effect must be accounted for. The transfer process depends on the magnification of the optical system, which determines the relationship between the coordinates in different spaces. The convolution between the PSF and the illumination pinhole in the spatial domain is given in Eq. (9.13), which accounts for both the PSF and the pinhole's effects on light propagation, yielding the desired intensity distribution.

Alternatively, convolution can be conducted in the frequency domain, where the PSF and the illumination pinhole function are multiplied. This allows for a convenient mathematical representation of the combined effect of the PSF and the pinhole on

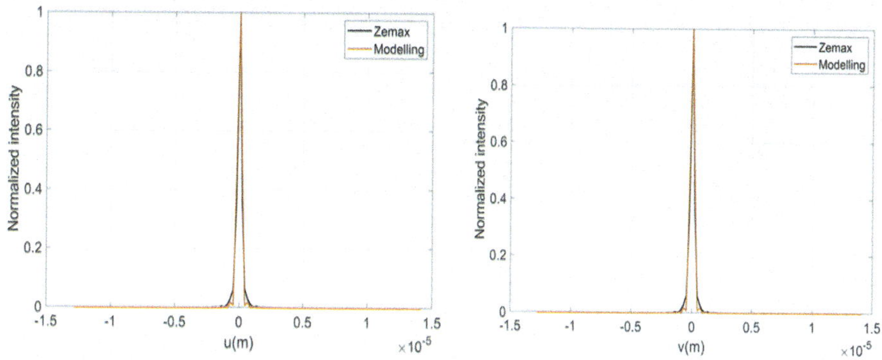

Fig. 9.4 Comparison of PSF cross section obtained from Zemax and proposed wave theory model

the light intensity distribution in the object space, as expressed in Eq. (9.14) [2].

$$I_{\mathrm{obj}}(x, y) = |h_{\mathrm{ill}}(x, y)|^2 \otimes_2 I_{\mathrm{ill}}(x, y) \tag{9.13}$$

$$I_{\mathrm{obj}}(x, y) = \mathcal{F}^{-1}\left(\mathcal{F}\left(|h_{\mathrm{ill}}(x, y)|^2\right) \cdot \mathcal{F}(I_{\mathrm{ill}}(x, y))\right) \tag{9.14}$$

where the symbol "$\otimes_2$" denotes convolutions in two-dimensional space, and the symbol "·" denotes element-wise multiplication. Figure 9.5 shows the pinhole and its convolution with PSF in object space.

Figure 9.6 shows a schematic of the object-detection space. Following reflection from the sample, light travels through the chromatic objective and passes through two achromatic lenses, which focus the light on the detector pinhole.

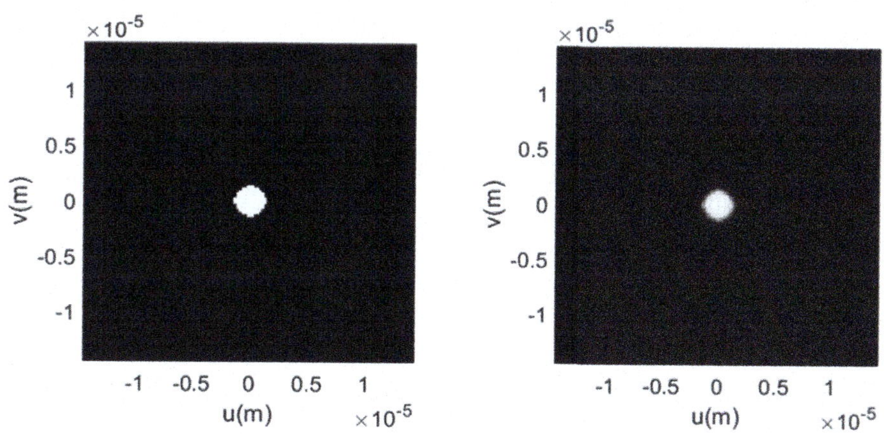

Fig. 9.5 Convolution result of illumination space pinhole in object space

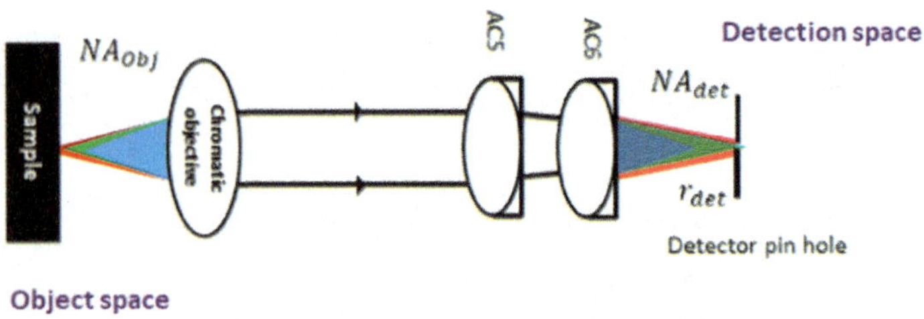

Fig. 9.6 Schematic of the object-detection space

Applying the same principle used to determine the PSF of the illumination-object space, the modeling of the object-detection space was done at a wavelength of 500.53 nm. The relevant modeling parameters can be found in Table 9.3.

The circular pupil function object-detection space of a diffraction-limited optical system is given by:

$$P_{\mathrm{img}}(x, y) = \mathrm{circ}\left(\frac{\sqrt{x^2 + y^2}}{R_{X\mathrm{Pimg}}}\right) \tag{9.15}$$

where $R_{X\mathrm{Pimg}}$ represents half of the exit pupil diameter of the object space, $D_{X\mathrm{Pimg}}$. The coherent transfer function can be expressed as follows:

$$H_{\mathrm{img}}(f_U, f_V) = \mathrm{circ}\left(\frac{\sqrt{(-\lambda Z_{X\mathrm{Pimg}} f_U)^2 + (-\lambda Z_{X\mathrm{Pimg}} f_V)^2}}{w_{X\mathrm{Pimg}}}\right) \tag{9.16}$$

Or

$$H_{\mathrm{img}}(f_U, f_V) = \mathrm{circ}\left(\frac{\sqrt{(f_U^2 + f_V^2)}}{f_0}\right) \tag{9.17}$$

where f_0 represents the coherent cutoff frequency, while f_U and f_V denote the frequency coordinates. f_0 is given as:

Table 9.3 Object-detection space parameters

Parameter	Values	Units
Wavelength (λ)	500.53	nm
Exit pupil diameter ($D_{X\mathrm{Pill}}$)	44.03103	mm
Exit pupil distance ($Z_{X\mathrm{Pill}}$)	389.9236	mm
Detector space width (square shape)	378.88	μm

$$f_0 = \frac{w_{XP\mathrm{img}}}{\lambda Z_{XP\mathrm{img}}} \tag{9.18}$$

The PSF or incoherent impulse response, represented by $|h|^2$, can be calculated using the following equation:

$$\mathrm{PSF}_{\mathrm{img}} = |h_{\mathrm{img}}(u, v)|^2 = |\mathcal{F}^{-1}\{H_{\mathrm{img}}(f_U, f_V)\}|^2 \tag{9.19}$$

Figure 9.7 illustrates the obtained diffraction-limited PSF in the detection space.

To validate the PSF obtained from the wave theory-based parametric model, it is compared with the PSF obtained from Zemax, as shown in Fig. 9.8. This comparison serves as a validation step to ensure the consistency between the theoretical predictions and the simulations conducted using Zemax.

The $\mathrm{PSF}_{\mathrm{img}}$ obtained from the model is convolved with the images acquired in the object space. This convolution operation combines the blurring effect of the PSF with the object space images to simulate the observed effect on the detection space, as described by Eqs. (9.20) and (9.21) in the spatial and frequency domain, respectively.

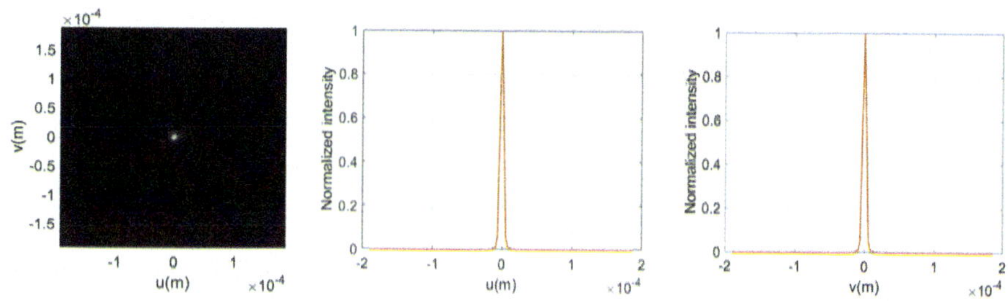

Fig. 9.7 Diffraction-limited imaging PSF and its cross section in the detection space

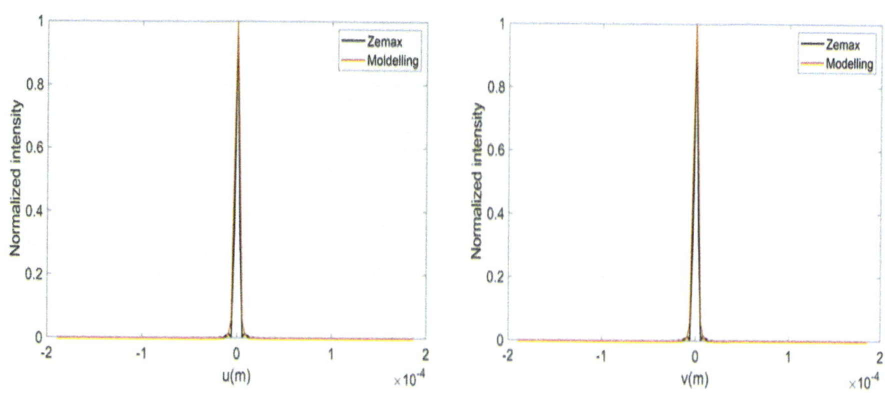

Fig. 9.8 Comparison plot of PSF cross section from Zemax and proposed wave theory model in detection space

$$I_{\text{img}}(x, y) = \left|h_{\text{img}}(x, y)\right|^2 \otimes_2 I_{\text{obj}}(x, y) \tag{9.20}$$

$$I_{\text{img}}(x, y) = \mathcal{F}^{-1}\left(\mathcal{F}\left(\left|h_{\text{img}}(x, y)\right|^2\right) \cdot \mathcal{F}\left(I_{\text{obj}}(x, y)\right)\right) \tag{9.21}$$

where h_{img} is the PSF obtained in detection space, I_{obj} is the intensity distribution of object space, and I_{img} is the resultant intensity distribution in detection space just before the pinhole.

Note that before convolving with the detection space PSF, it is necessary to transfer the convolved image from the object space to the detection space. By correctly transferring the image to the detection space, we can perform the final convolution operation with the detector space PSF to obtain the simulated image, as depicted in Fig. 9.9.

Determining the pinhole size in the detection space depends on the magnification of the optical system, as provided in Table 9.1. Given that the illumination pinhole size used in the system is 30 μm, the optimized pinhole size used in the detector space is determined to be 50 μm. The pinhole in the detection space plane selectively filters out defocused light from wavelengths other than the intended wavelength. The equation of the filtered light field is given as:

$$I_{\text{detector}}(x, y) = I_{\text{img}}(x, y) \cdot D_{\text{s}}(x, y) \tag{9.22}$$

where $I_{\text{detector}}(x, y)$ is the intensity distribution of the detected image just after the pinhole at coordinates (x, y), $I_{\text{img}}(x, y)$ is the intensity distribution of the original image just before the pinhole at the same coordinates, and $D_{\text{S}}(x, y)$ is the mask or

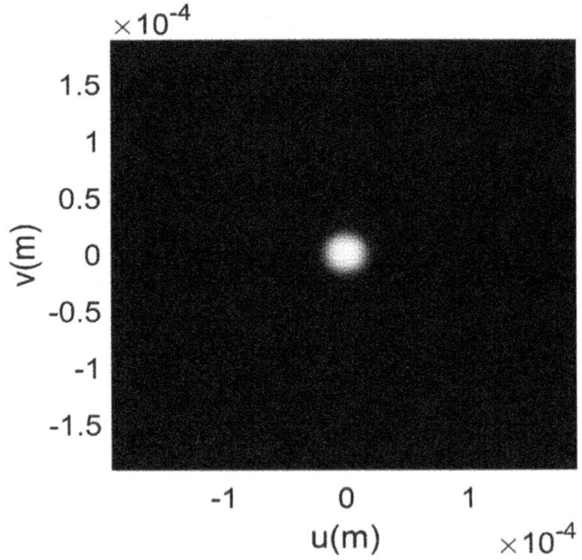

Fig. 9.9 Spot in detection plane

filter applied by the pinhole at coordinates (x, y). The symbol "·" denotes the element-wise multiplication between the image and the pinhole mask. The spot image filtered by the detection pinhole is shown in Fig. 9.10.

The spectrometer system comprises several essential components, including two achromatic lenses, a diffraction grating, and a camera. These components work together to capture and analyze the spectrum of incoming light. Figure 9.11 shows the schematic of the spectrometer.

Following the same principle discussed before for PSF modeling, the modeling of the camera-detection space was done. The relevant modeling parameters can be found in Table 9.4.

The circular pupil function for camera-detection space of a diffraction-limited optical system is given by:

$$P_{\text{spectrometer}}(x, y) = \text{circ}\left(\frac{\sqrt{x^2 + y^2}}{R_{XP\text{spectrometer}}}\right) \tag{9.23}$$

where $R_{XP\text{spectrometer}}$ represents half of the exit pupil diameter of the object space, $D_{XP\text{spectrometer}}$. The coherent transfer function can be expressed as follows:

$$H_{\text{spectrometer}}(f_U, f_V) = \text{circ}\left(\frac{\sqrt{\left(-\lambda Z_{XP\text{spect}} f_U\right)^2 + \left(-\lambda Z_{XP\text{spect}} f_V\right)^2}}{w_{XP\text{spectrometer}}}\right) \tag{9.24}$$

Or

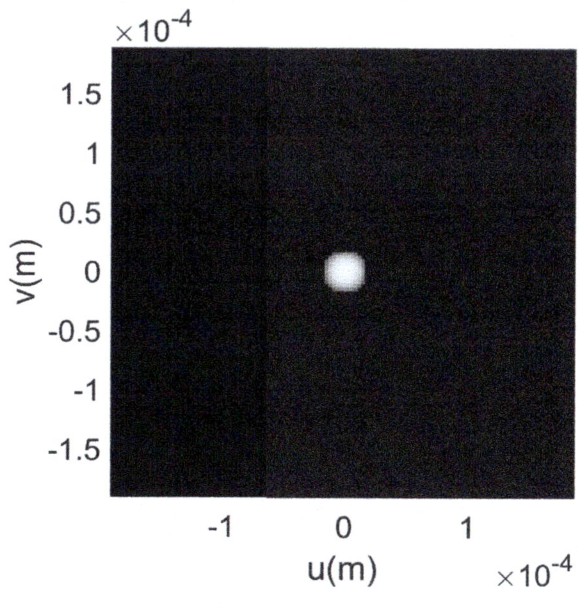

Fig. 9.10 Filtered spot at the detection pinhole

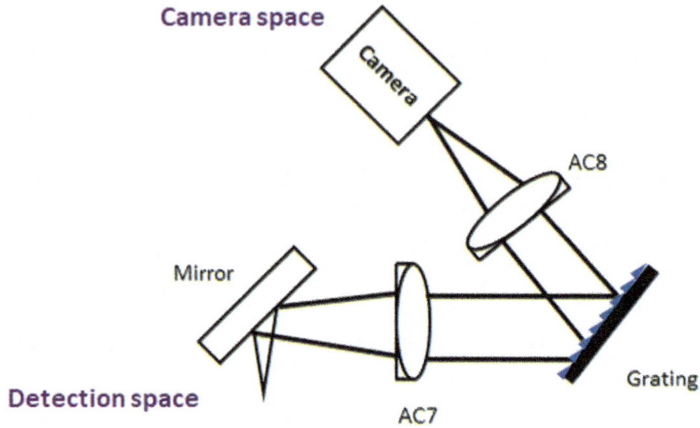

Fig. 9.11 Schematic of the spectrometer system (detection-camera space)

Table 9.4 Parameters of the camera-detection space

Parameter	Values	Units
Wavelength (λ)	500.53	nm
Exit pupil diameter ($D_{XP\text{spectrometer}}$)	59.80702	mm
Exit pupil distance ($Z_{XP\text{spectrometer}}$)	362.6096	mm
Detector space width (square shape)	236.8	μm

$$H_{\text{spectrometer}}(f_u, f_v) = \text{circ}\left(\frac{\sqrt{(f_U^2 + f_V^2)}}{f_0}\right) \tag{9.25}$$

where f_0 represents the coherent cutoff frequency, while f_U and f_V denote the frequency coordinates. f_0 is given as:

$$f_0 = \frac{R_{XP\text{spectrometer}}}{\lambda Z_{XP\text{spectrometer}}} \tag{9.26}$$

The PSF or incoherent impulse response, represented by $|h|^2$, can be calculated by:

$$PSF_{\text{spectrometer}} = \left|h_{\text{spectrometer}}(u, v)\right|^2 = \left|\mathcal{F}^{-1}\left\{H_{\text{spectrometer}}(f_U, f_V)\right\}\right|^2 \tag{9.27}$$

The obtained diffraction-limited PSF is shown in Fig. 9.12.

To validate the obtained PSF from the wave theory-based parametric model for the camera-detection space, it is compared with the PSF obtained from Zemax, as shown

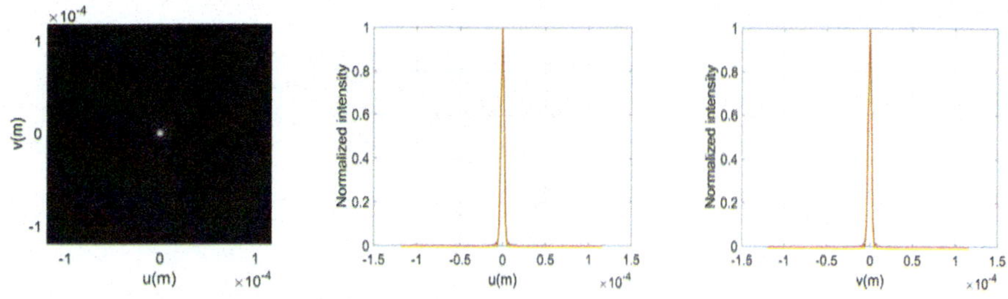

Fig. 9.12 PSF of spectrometer in camera space

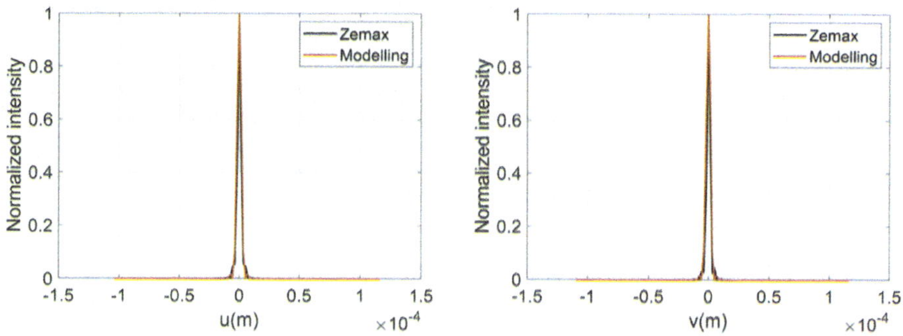

Fig. 9.13 Comparison of PSF obtained from Zemax and the proposed wave theory model in the camera space

in Fig. 9.13. This comparison serves as a validation step to ensure the consistency between the theoretical predictions and the simulations conducted using Zemax.

To obtain the final spectrum at the camera plane, the PSF obtained from the modeling process is convolved with the filtered image captured at the detection plane. By correctly transferring the image to the detection space, we can perform the final convolution operation with the detector space PSF to obtain the simulated image, as described by Eqs. (9.28) and (9.29) in the spatial and frequency domain.

$$I_{\text{spectrometer.}}(x, y) = \left|h_{\text{spectrometer}}(x, y)\right|^2 \otimes_2 I_{\text{detector}}(x, y) \tag{9.28}$$

$$I_{\text{spectrometer}}(x, y) = \mathcal{F}^{-1}\left(\mathcal{F}\left(\left|h_{\text{spectrometer}}(x, y)\right|^2\right) \cdot \mathcal{F}(I_{\text{detector}}(x, y))\right) \tag{9.29}$$

where $h_{\text{spectrometer}}$ is the PSF obtained in camera space, I_{detector} is the intensity distribution of filtered intensity, and $I_{\text{spectrometer}}$ is the resultant intensity distribution in camera space.

The obtained spectrum at the camera plane is shown in Fig. 9.14.

To determine the PSF of the entire system, the PSFs of the illumination, imaging, and spectrometer components are convolved together. The resulting PSF describes

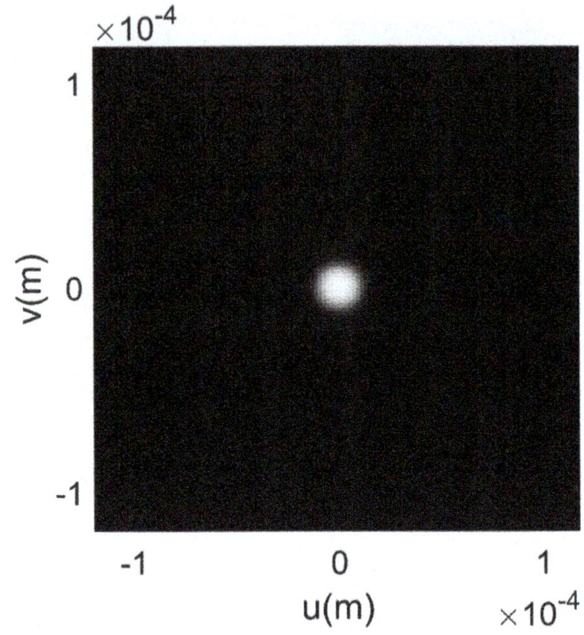

Fig. 9.14 Resultant spectrum at the camera space

the blurring and spreading of light caused by the combined effect of these components. The resultant PSF, along with its cross section, is shown in Fig. 9.15. The resultant PSF for the reflection part is as follows [3].

$$PSF_{\text{resultant}} = PSF_{\text{img}} \otimes_2 PSF_{\text{spectrometer}} \tag{9.30}$$

Applying the identical approach, the modeling of the spectrum for various wavelengths has been derived. Figure 9.16 shows the relationship between the axial response signal and the confocal spectral signal. To establish a model for the 2-D spectral signal, an axial confocal signal model with a monochromatic source

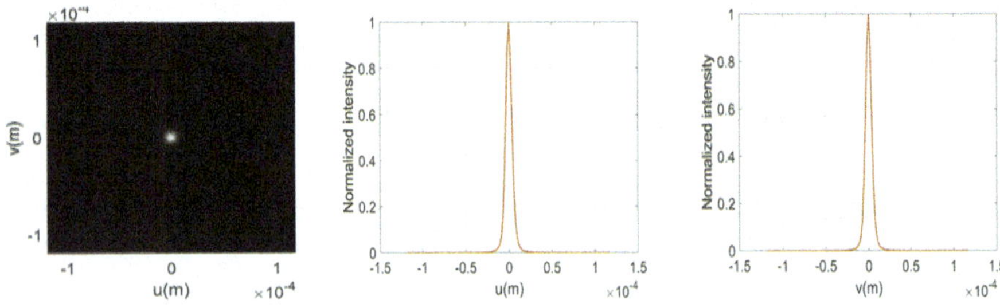

Fig. 9.15 Resultant PSF and its cross-section view in the camera space

is created. Additionally, the calculation of the axial displacement-wavelength relationship is crucial, achieved through experimentally acquiring the depth response curve.

However, it is important to note that real optical elements are never ideal and are subjected to aberrations. Aberrations introduce deviations from the ideal behavior, and their impact varies across the image field. To realize a more realistic representation of the spectrum signal, it is necessary to incorporate the effect of aberrations into the obtained spectrum.

Figure 9.17 compares the experimental and simulated signals affected by aberrations. The experimental signal, which will be discussed in detail in the next chapter, provides further insights into the actual captured spectrum. By considering the effects of aberrations and incorporating experimental findings, the obtained spectrum can better reflect real-world conditions.

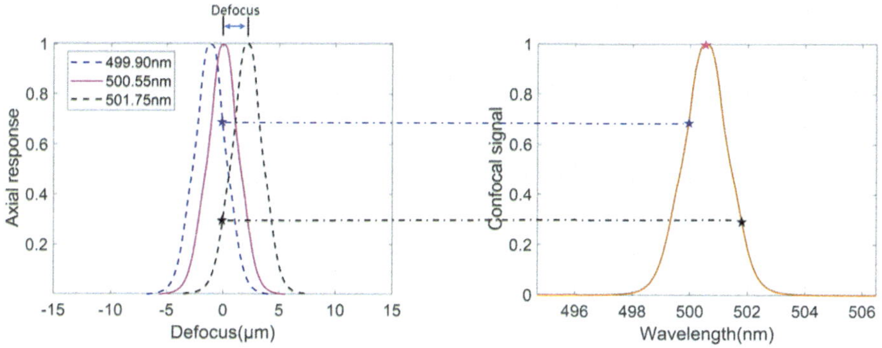

Fig. 9.16 The relationship between the confocal spectral signal and axial response signal

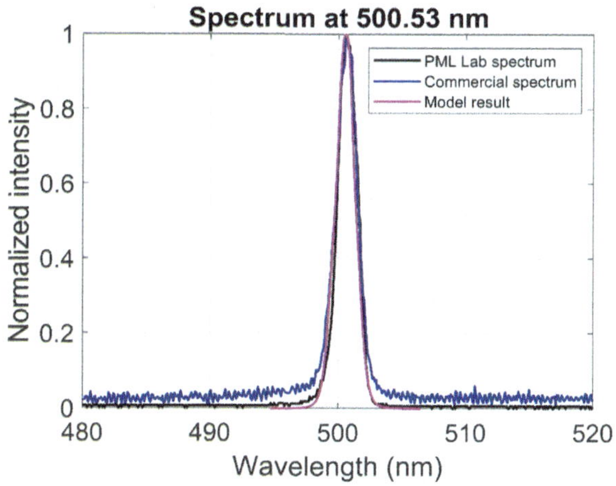

Fig. 9.17 Comparison of experimental and model outcome spectrum at 500.53 nm

Comparisons at other wavelengths were conducted, as shown in Figs. 9.18 and 9.19. Table 9.5 shows the FWHM values of these spectrums and their deviations from the result measured by a commercial spectrometer. A noticeable difference between the FWHM values obtained from the model and the experiment is observed. This difference is attributed to the absence of noise consideration in the model, as it can affect the FWHM of the spectrum [4].

The system employs a chromatic objective featuring a dispersion wavelength range from 450 to 650 nm, covering a depth range of 300 μm. Various parameters such as wavelength, magnification, exit pupil, and entrance pupil diameter vary across the focal range. With careful consideration of these parameters, the spectral signal across diverse positions has been comprehensively modeled. The resultant representation is shown in Fig. 9.20, alongside the depth response of the confocal signal.

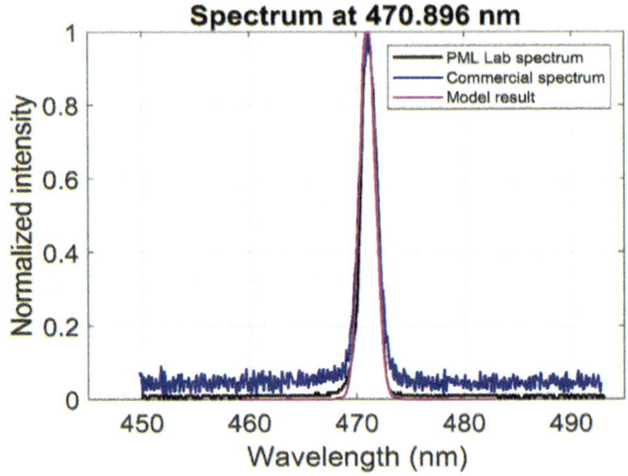

Fig. 9.18 Comparison of experimental and model outcome spectrum at 470.896 nm

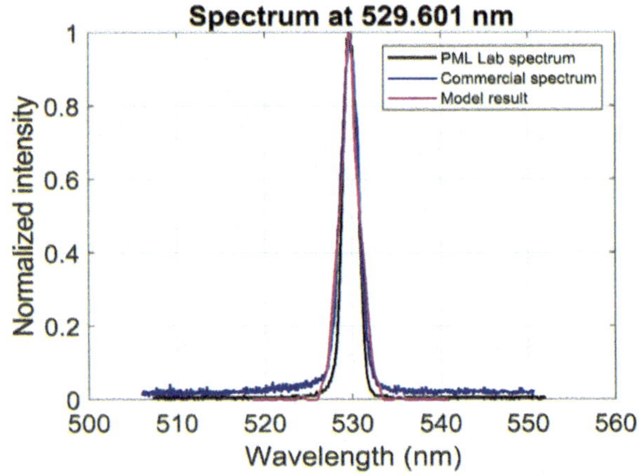

Fig. 9.19 Comparison of experimental and model outcome spectrum at 529.601 nm

Table 9.5 Comparison of commercial, PML Lab in NTU, and model results

		FWHM (nm)	Deviation (nm)	% Deviation
Spectrum at 470.896 nm	Commercial	1.86		
	PML Lab	1.72	0.14	7.52
	Model result	1.74	0.11	6.45
Spectrum at 500.53 nm	Commercial	1.95		
	PML Lab	1.88	0.06	3.723
	Model result	1.76	0.19	9.74
Spectrum at 529.601 nm	Commercial	2.398		
	PML Lab	2.052	0.34	14.42
	Model result	2.3589	0.03	4.8

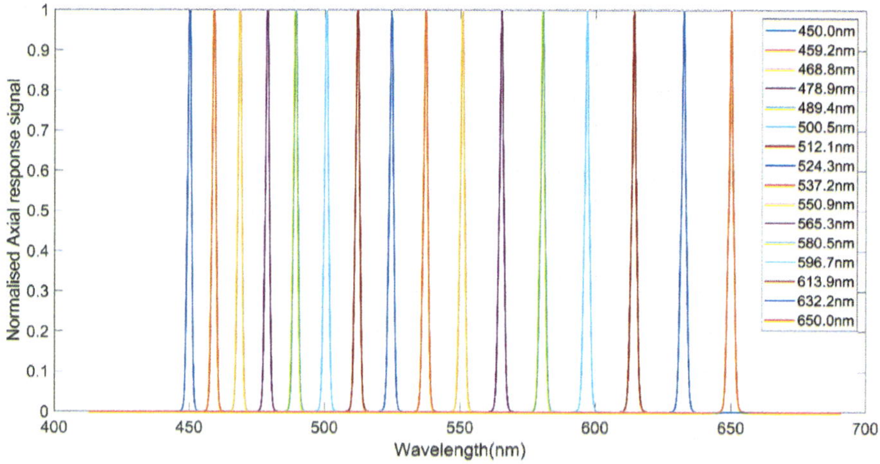

Fig. 9.20 The modeled confocal spectral signal at different dispersion positions

9.3 Weiner Deconvolution by Deep Learning

Weiner deconvolution [5] is a common technique employed to recover a degraded or blurred signal by estimating the original signal that might have produced it. A key element in Weiner deconvolution is the PSF, which describes the blurring effect or degradation introduced to the signal. The PSF represents the system's response to a point source and is an essential component in the deconvolution process.

The traditional Weiner deconvolution formula incorporates several parameters to estimate the original signal from a degraded or blurred signal. The formula can be expressed as:

$$\hat{X}(u, v) = \frac{H^*(u, v)Y(u, v)}{|H(u, v)|^2 + \frac{1}{\mathrm{SNR}(u,v)}} \tag{9.31}$$

where (u, v) is the frequency space coordinate, $\hat{X}(u, v)$ is the estimated frequency response of the original signal, $H^*(u, v)$ denotes the complex conjugate of the PSF, and $Y(u, v)$ represents the observed degraded image in the frequency domain. Additionally, $|H(u, v)|^2$ represents the squared magnitude PSF and $\mathrm{SNR}(u, v)$ denotes the signal-to-noise ratio of the measurement.

The traditional Weiner deconvolution aims to minimize the mean square error between the estimated signal, $X(u, v)$, and the original signal. This is achieved by applying a frequency-dependent weighting $Y(u, v)$ to the degraded signal's frequency response and using the inverse of the degradation function's frequency response magnitude squared $|H(u, v)|^2$, along with the signal-to-noise ratio term $1/\mathrm{SNR}(u, v)$.

The signal-to-noise ratio term, $1/\mathrm{SNR}(u, v)$, serves as a regularization or smoothing factor. It adds a small value to the denominator of the formula, ensuring that the resulting estimate of the original signal is not overly sensitive to noise in regions where the signal-to-noise ratio is low. This addition helps prevent excessive amplification of noise during the deconvolution process.

By adjusting the $\mathrm{SNR}(u, v)$ parameter, the traditional Weiner deconvolution formula allows for a trade-off between signal fidelity and noise amplification. Higher values of $\mathrm{SNR}(u, v)$ prioritize noise suppression but may result in a loss of fine details in the estimated signal. Conversely, lower values of $\mathrm{SNR}(u, v)$ preserve more details but may be susceptible to noise amplification. The choice of an appropriate regularization parameter depends on the specific characteristics of the degraded signal and the noise present in the system. It is important to find a balance between suppressing noise while retaining key signal characteristics.

It's worth noting that regularization can be incorporated into the deep learning framework. Techniques such as weight decay or dropout can be applied during deep neural network training to introduce regularization and prevent overfitting to the training dataset.

Deep Weiner deconvolution [6, 7] extends the traditional Weiner deconvolution approach by incorporating deep learning techniques. Instead of relying solely on predefined mathematical models and assumptions, deep Weiner deconvolution harnesses deep neural networks to learn the mapping between the degraded signal and the original signal, considering the characteristics of the PSF.

The PSF's importance lies in its ability to capture the system's blurring characteristics. Traditional Weiner deconvolution assumes linearity and stationarity, relying on an estimated PSF that is typically known or estimated from the degradation process. However, real-world scenarios often involve complex and non-linear blurring effects that cannot be accurately modeled by simple mathematical functions. Deep neural networks, as utilized in deep Weiner deconvolution, have the capability to capture and learn intricate and non-linear relationships, including the complex behavior of the PSF.

By training a deep neural network on a large dataset containing paired degraded and original signals, the network not only learns the mapping between the degraded and original domains in a general sense but also adapts to the specific characteristics of the PSF. The network learns to extract relevant features and patterns from the input signal while considering the PSF's impact on the degradation process. This adaptability allows deep Weiner deconvolution to effectively restore degraded signals, even when confronted with unknown or complex PSFs. The practical realization of deep Weiner deconvolution will be explained in the following subsections.

9.3.1 Preparation of Dataset

Three data types are required to train the deep learning model for Weiner deconvolution: input images, corresponding PSFs, and corresponding ground truth data. In the previous section, a parametric model based on scalar diffraction wave theory was developed. This model enables one to generate input images for specific heights, along with their corresponding PSFs and ground truth data.

The input image represents the spectrum image captured by the camera, while the PSF corresponds to the optical system's PSF in both the object-detection space and detection-camera space. The ground truth data represents the signal in the object space, obtained through the convolution of the illumination space PSF and the illumination pinhole. All these images have dimensions of 512×256 pixels.

The dataset consists of nine different depths: 99.687, 100, 100.8815, 100.66, 101.5426, 104.95, 105, 105.987, and 110 μm. Among these, spectrum images for 100, 100.8815, 105, and 110 μm are generated using the parametric model. The spectrum images for the depths of 99.687, 100.66, 101.5426, 104.95, and 105.987 μm are obtained from experimental measurements. Each depth has 80 spectrum images, and for the experimentally obtained spectrum, different images are captured at different times for each depth. Therefore, the total number of input images is 720. For each depth, there is a corresponding PSF and ground truth image, resulting in a total of nine PSFs and nine ground truth images. A block diagram illustrating the dataset directory path is provided in Fig. 9.21.

To train the model, 80% of the total dataset (576 images) was used. The remaining 20% (144 images) are reserved for testing the model's performance, as depicted in Fig. 9.22. This split allows for the evaluation of the model's ability to generalize and perform well on unseen data.

By utilizing the proposed training approach, the deep learning model can learn to perform Wiener deconvolution by estimating the underlying ground truth signal from input images, guided by the corresponding PSFs. Subsequently, the model's capacity to generalize and deconvolve new input data can be evaluated during the testing phase with the held-out images.

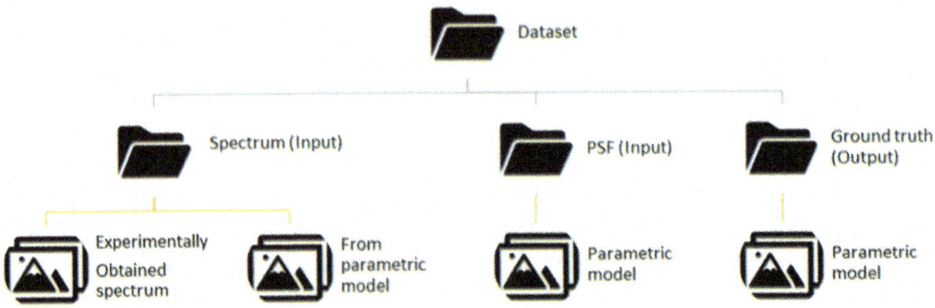

Fig. 9.21 Example folder directory of the dataset

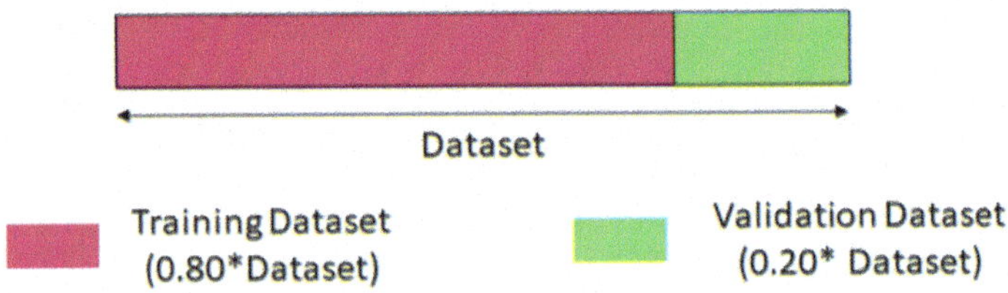

Fig. 9.22 Split arrangement of dataset

9.3.2 *Weiner Deconvolution by Deep Learning*

The block diagram of deep learning-based Wiener deconvolution is shown in Fig. 9.23. As shown in the figure, the deep learning deconvolution process consists of two main components. Initially, Wiener deconvolution is applied to each input spectrum, $Y_k(u, v)$, with Point Spread Functions (PSF), denoted as $(H_i(u, v))$ in the frequency domain. The variable "k" ranges from 1 to the total number of input spectrum images, and "i" ranges from 1 to the total number of PSFs.

Upon completing the deconvolution, an inverse Fourier transform is performed to convert the data back into the spatial domain. These deconvolved images are then input into the deep learning model, where the model learns from the most accurately deconvolved spectrum. The model comprises an encoder and decoder linked by skip connections. The difference between the ground truth and the model's output is calculated using the Huber loss function.

The code consists of two main classes: Weiner deconvolution and an encoder-decoder structure with skip connections. Note that the encoder-decoder structure can also be represented as a U-net.

The Wiener deconvolution class is a PyTorch module that implements the Wiener deconvolution algorithm. It takes an initial PSF as input and performs Wiener deconvolution by calculating the Fourier transform of the input image, multiplying it with the complex conjugate of the PSF, and dividing by the magnitude squared of the PSF plus a term proportional to the signal-to-noise ratio (SNR). The result is then inverse Fourier transformed and returned as the output.

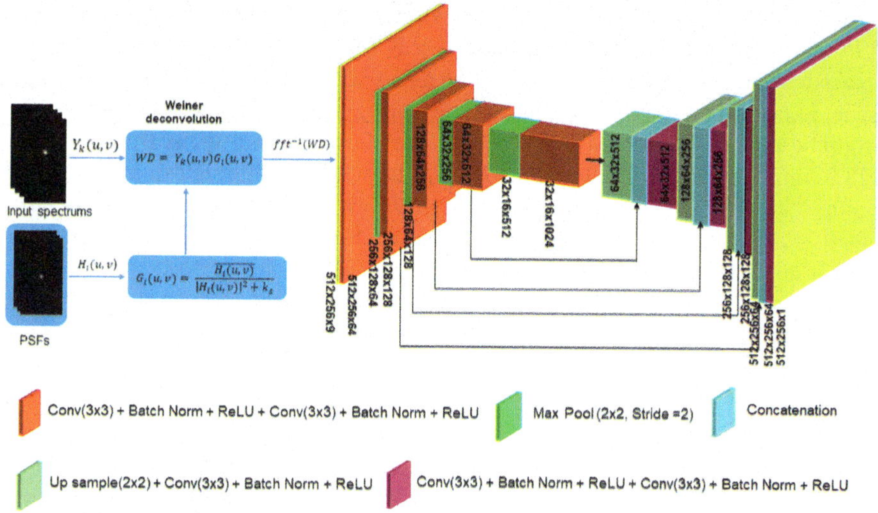

Fig. 9.23 Deep learning-based Weiner deconvolution model architecture

The encoder-decoder class architecture with skip connections is commonly used for image segmentation tasks. It consists of multiple convolutional blocks and up-convolution blocks to capture and upsample features at different resolutions. The input image passes through these blocks, with skip connections combining features from earlier layers with those from later layers. The output is obtained after a final convolutional layer and a sigmoid activation function. Note that skip connections are required in large models to preserve the information of the layers, ensuring that crucial details are not lost during processing.

9.3.3 Size of the Input Spectrum, PSF, and Output Spectrum

In this example, nine point spread functions (PSFs) are utilized. Each input spectrum will be deconvolved with these nine PSFs, resulting in nine input spectra for the deep learning model. Consequently, the input size will be (512 × 256 × 9). This step is necessary because sharp changes, such as edges in step height or bump samples, can cause input spectra to encompass multiple depth positions rather than just one. Therefore, it is not ideal to deconvolve the input spectrum with only one PSF. By using PSFs corresponding to different height positions, the input spectrum is deconvolved, and the resulting nine deconvolved images are fed into the encoder-decoder model with skip connections. The model learns to identify the spatial positions that are closer to the ground truth among these nine deconvolved images. The output from the deep learning model is a single image of size 512 × 256.

To train the deep learning model for Wiener deconvolution, the input spectrum, PSF, and output spectrum are all of size 512 × 256 pixels. However, the total spectrum size extends over 2087 pixels, resulting in a large dataset for training. To handle this issue, a specific region of the spectrum measuring 512 × 256 pixels is cropped. The cropping region is depicted in Fig. 9.24. The chosen region spans from 481.53 to 528.93 nm, corresponding to 512 pixels in the wavelength dimension. In terms of depth, it covers 82.11 μm. The plot data of the cropped region can be observed in Fig. 9.24. During sample measurement, the object space is adjusted so that the entire sample depth falls within these cropping regions.

Each input spectrum will have a corresponding ground truth spectrum value in the output. The hyperparameters related to the training and testing of the model are provided in Table 9.6.

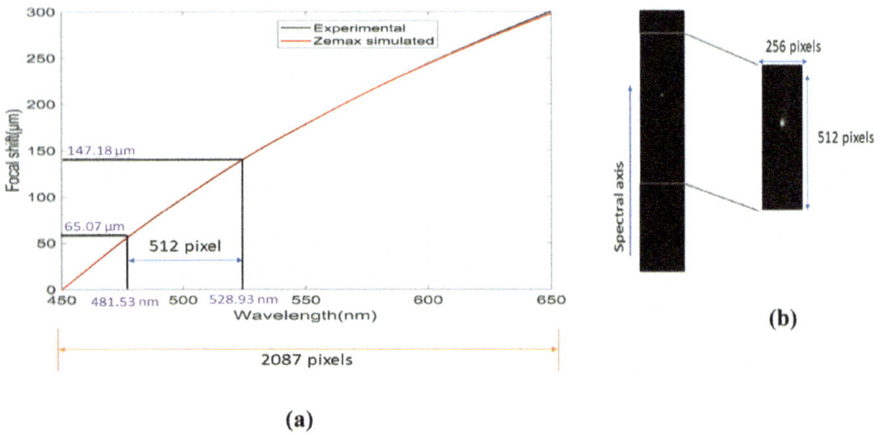

Fig. 9.24 **a** Plot showing the cropping region of the spectrum and **b** the cropped spectrum

Table 9.6 Description of hyperparameters

Parameter name	
Learning rate	0.7e−4
Loss function	Huber loss
Optimization name	ADAM
Activation function	Sigmoid
Batch size	15
Epochs	600
Input spectrum	720
PSFs	9

9.3.4 Hyperparameters

Continuing the example mentioned in the previous section, a batch size of 15 was chosen during the training process. The batch size refers to the quantity of samples handled in each iteration or update of the model's parameters.

The training data is divided into smaller groups or mini-batches, where the model updates its weights based on the average gradient computed from these mini-batches. This approach offers several advantages, such as faster convergence, reduced memory requirements, and improved generalization due to the inclusion of diverse samples in each batch. The batch size is chosen to find a balance between computational efficiency and effective learning. It ensures that the model receives an adequate number of samples in each iteration to capture meaningful patterns while still maintaining computational feasibility. The training set, comprising 80% of the total dataset, is used to update the model's parameters iteratively through a specified number of epochs. During each epoch, the model processes the training data in multiple batches of size 15, updating its weights based on the calculated gradients and using the ADAM optimization algorithm. The remaining 20% of the dataset forms the testing set, which is held separately from the training process. After the model is trained, it is evaluated on the independent testing set to assess its performance on unseen data. This evaluation provides valuable insights into the model's ability to generalize and make accurate predictions on new spectra.

As previously mentioned, the ADAM optimization algorithm is employed to train the deep learning model. ADAM, short for Adaptive Moment Estimation, is commonly employed in deep learning due to its effectiveness in optimizing model parameters. It is an adaptive learning rate optimization algorithm that merges the benefits of two other well-known optimization techniques: Adaptive Gradient Algorithm (AdaGrad) and Root-Mean-Square Propagation (RMSProp).

ADAM keeps an exponentially weighted moving average of previous gradients and squared gradients. This enables it to dynamically adjust the learning rate based on the historical magnitudes of gradients, promoting faster convergence and improved training efficiency. Additionally, ADAM incorporates bias correction to address the biases introduced during the initialization phase.

With ADAM optimization, the model can efficiently update its parameters during training to minimize the loss function and improve the overall performance. It helps to mitigate issues such as vanishing or exploding gradients and provides a robust and reliable optimization framework for training deep neural networks.

Moreover, the Huber loss function was employed to train the deep learning model. Frequently applied in regression tasks, the Huber loss is particularly useful when handling datasets containing outliers or noisy data points. It offers a balance between the Mean-Squared Error (MSE) loss and the Mean Absolute Error (MAE) loss, making it a suitable choice for scenarios where both accuracy and robustness are desired.

The Huber loss combines the benefits of the MSE loss, which penalizes larger errors more severely, and the MAE loss, which provides a more robust estimation

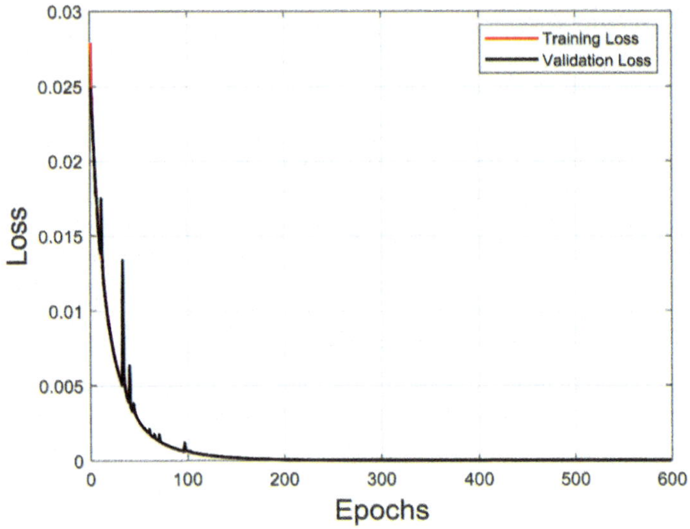

Fig. 9.25 Plot of training and validation loss

by penalizing errors linearly. This combination allows the Huber loss to handle outliers or noisy data points while still considering the overall data trend. The Huber loss function is defined by a threshold parameter called "delta". When the absolute difference between the predicted and target values is smaller than the delta, the loss is computed as the squared error, resembling the MSE loss. Conversely, when the absolute difference exceeds the delta, the loss is computed as the linear difference, resembling the MAE loss. The threshold delta determines the point at which the loss transitions from quadratic to linear.

By utilizing the Huber loss, the deep learning model becomes more resilient to outliers, ensuring that their influence on the training process is appropriately attenuated. This allows the model to focus on accurately capturing the patterns exhibited by the majority of the data while still accounting for larger errors. The description of hypermeters used in the proposed deep learning model is given in Table 9.6.

The plot for training and testing loss is shown in Fig. 9.25, which contains the loss for 600 epochs.

From Fig. 9.25, it is observed that the testing and testing loss approach zero after 200 epochs. To evaluate the experimental data, the epoch at 350 was chosen as it converges and yields better results for all experimentally input spectra.

9.3.5 Deconvolution Results on Testing Data

To evaluate the performance of the trained model, unseen data was used for testing. The testing sample consists of a spectrum with a peak wavelength of 506.793 nm,

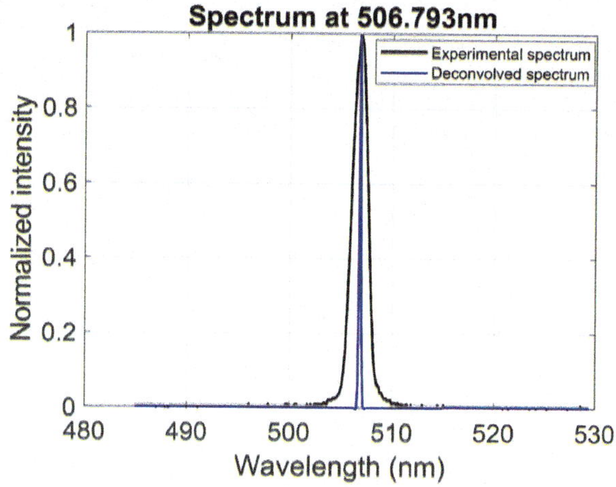

Fig. 9.26 Comparison of the effect of deconvolution for a spectrum at a wavelength of 506.793 nm

Table 9.7 Effect of deep learning-based deconvolution on FWHM

	FWHM (nm)
Without deconvolution	1.7667
With deconvolution	0.157

which was absent in the training dataset. This ensures that the model is tested on data it has not encountered previously, enabling a fair assessment of its generalization capabilities.

The deconvolution effect of the model can be observed by analyzing the resulting images given in Fig. 9.26. The figure visually demonstrates the impact of the deconvolution process, providing a qualitative comparison between the original spectrum and the deconvolved spectrum. This analysis allows for the assessment of the effectiveness of deconvolution in removing defocused light and enhancing the resolution of the system.

Additionally, Table 9.7 provides quantitative results regarding the FWHM, indicating a reduction of approximately 11 times in FWHM. The FWHM reduction reflects the extent to which defocused light has been removed through the deconvolution process. As a greater reduction in FWHM signifies a more significant improvement in the system's resolution, Table 9.7 also provides a quantitative measure of deconvolution's effectiveness in enhancing the resolution of the system.

The effectiveness of the proposed deep learning-based deconvolution was also assessed on the 1.08 μm step height spectrum, as shown in Fig. 9.27 Effect of deep learning-based deconvolution on step height spectrum. It is observed from the figure that the developed convolution process significantly enhances the sharpness of the spectrum distribution, reducing the FWHM from 1.76 to 0.15 nm.

In summary, the testing phase of the model sheds light on the effects of deconvolution on the spectral data. Qualitative and quantitative analysis demonstrates the model's ability to improve the system's resolution and eliminate defocused light.

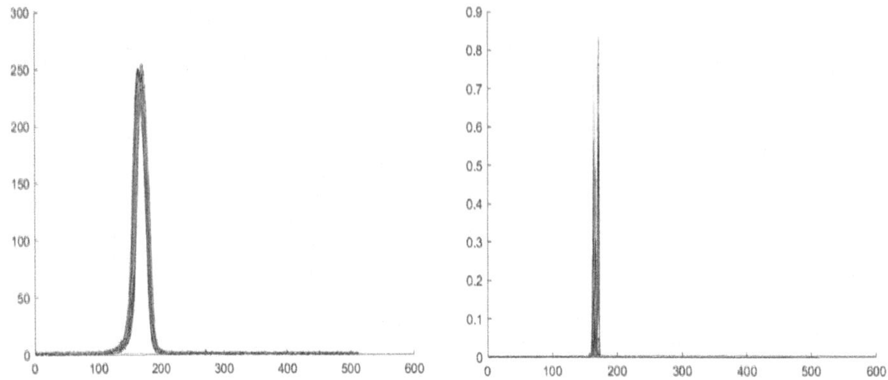

(a) Before deconvolution (b) After deconvolution

Fig. 9.27 Effect of deep learning-based deconvolution on step height spectrum

References

1. Singh SK (2023) Theory and implementation of microscopic diffractive image profilometry. Master thesis, National Taiwan University
2. Goodman JW (2005) Introduction to Fourier optics. Roberts and Company Publishers
3. Olsovsky C, Shelton R, Carrasco-Zevallos O, Applegate BE, Maitland KC (2013) Chromatic confocal microscopy for multi-depth imaging of epithelial tissue. Biomed Opt Express 4(5):732–740
4. Chen L, Cottrell CE, Marshall AG (1986) Effect of signal-to-noise ratio and number of data points upon precision in measurement of peak amplitude, position and width in Fourier transform spectrometry. Chemom Intell Lab Syst 1(1):51–58
5. Chen L-C, Lin T-Y, Chang Y-W, Lin S-T (2013) Chromatic confocal surface profilometry employing signal recovering methodology for simultaneously resolving lateral and axial cross talk problems. In: Eighth international symposium on precision engineering measurement and instrumentation, 2013, vol 8759. SPIE, pp 1133–1138
6. Dong J, Roth S, Schiele B (2020) Deep wiener deconvolution: Wiener meets deep learning for image deblurring. Adv Neural Inf Process Syst 33:1048–1059
7. Yanny K, Monakhova K, Shuai RW, Waller L (2022) Deep learning for fast spatially varying deconvolution. Optica 9(1):96–99

Chapter 10
Practical Implementation of Chromatic Confocal Diffractive Image Profilometry (CC-DIP)

10.1 Introduction

This chapter will present the system configuration and the experimental setup of chromatic confocal diffractive image profilometry (CC-DIP). Furthermore, practices for real measurement, such as peak detection and depth calibration, are explained.

10.2 System Design

The system architecture is shown in Fig. 10.1, with detailed information given in Table 10.1. The system is divided into three modules: the illumination module, the chromatic confocal module, and the sensing module. The illumination module comprises the portion before the illumination space. The chromatic confocal module includes the portion of illumination-object-detection spaces. The sensing module consists of the camera space, which houses the spectrometer.

The chromatic confocal microscopy technique encodes depth information based on different wavelengths. Therefore, it requires a broadband light source, such as a Xenon lamp, a white light-emitting diode (LED), or a supercontinuum laser.

To integrate these three modules into a complete optical system, the design must ensure numerical aperture matching between modules to optimize light source usage and achieve optimal optical resolution. This indicates that the entrance and exit pupils of different modules must be properly matched. If the exit pupil of the illumination module is larger than the entrance pupil of the chromatic confocal module, light source utilization will decrease, and stray light will be introduced into the system. Conversely, if the exit pupil of the illumination module is smaller than the entrance pupil of the chromatic confocal module, the pupil of the chromatic objective will not be adequately filled, negatively impacting its optical performance and the overall optical lateral resolution.

L. Chen et al., *Diffractive Image Microscopy for 3D Imaging*, Springer Tracts in Mechanical Engineering, https://doi.org/10.1007/978-981-97-7782-2_10

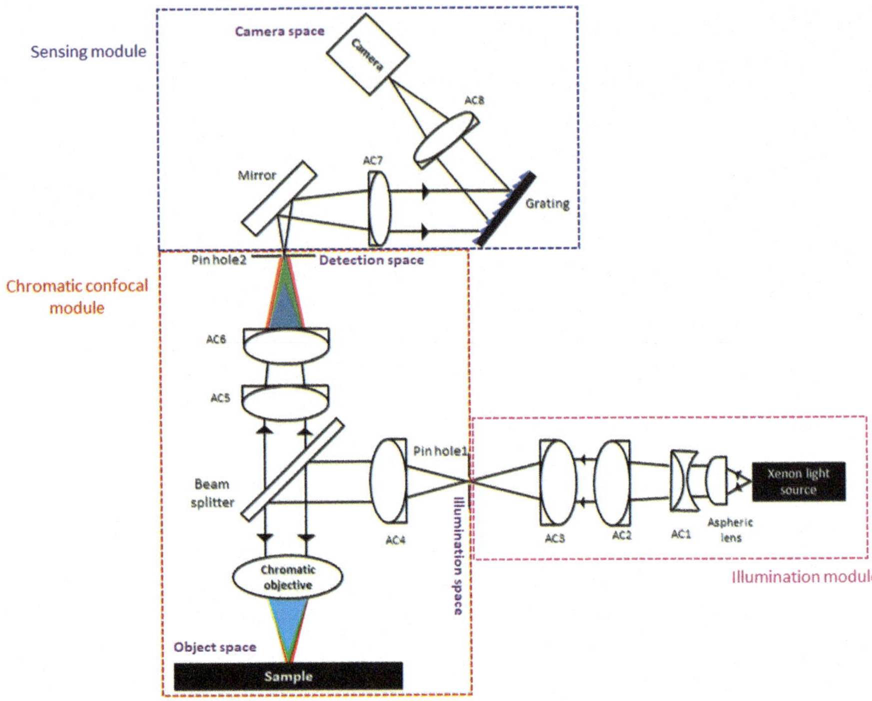

Fig. 10.1 Schematic of chromatic confocal microscopy

In this system, a 100-W Xenon lamp serves as the broadband light source, paired with a linear head fiber optic cable to create a preliminary linear light source. A high NA aspheric lens collects the light from the cable. Subsequently, three achromatic lenses are used to collimate and focus the light into the illumination space, forming the illumination module of the system.

The chromatic confocal module processes light emitted by the illumination module by first filtering through the illumination pinhole and then collimating it with an achromatic lens. A beam splitter reflects the light and directs it to the infinity-corrected chromatic objective, which focuses the light at different focal positions depending on its wavelength.

The light reflected from the sample is collected by the infinity-corrected chromatic objective and passes through the beam splitter to form another optical path. The light is then focused by the tube lens (consisting of two achromatic lenses) into the detection space, where a second pinhole is placed for additional spatial filtering. It then enters the sensing module, where an in-house designed spectrometer separates it into its spectral components. Due to the unique design of the chromatic objective, different wavelengths are focused at different depth planes, allowing only specific wavelengths to pass through the optical slit from the light reflected by the sample surface. By analyzing the reflected light spectrum, the associated depth information

Table 10.1 Prescription of optical elements used

System elements	Specification	Model name
Light source	Broadband halogen light source (NA—0.5877)	Hamamatsu, model: L9588-03
Aspheric lens	NA-0.67	Edmund#65-996
Achromatic doublet (AC1)	EFL = − 40 mm, clear aperture-25.4mm	ACN254-040-A
Achromatic doublet (AC2)	EFL = 150 mm, clear aperture-40 mm	DLB-40-150PM
Achromatic doublet (AC3)	EFL = 200 mm, clear aperture-40 mm	Edmund#32-925
Pinhole 1	30μm	Edmund#84-061
Achromatic doublet (AC4)	EFL = 160 mm, clear aperture-40 mm	Edmund #32-923
Beam splitter	50 mm	BS031
Chromatic objective	NA-0.72(object plane)	PML Lab
Achromatic doublet (AC5)	EFL = 400 mm, clear aperture-50 mm	ACT508-400-A-ML
Achromatic doublet (AC6)	EFL = 750 mm, clear aperture-50 mm	ACT508-750-A-ML
Pinhole 2	50 μm	SPHP-50
Achromatic doublet (AC7)	EFL = 200 mm, clear aperture- 30mm	Edmund #47-719
Reflective diffraction grating	150line/mm	Edmund #64-402
Achromatic doublet (AC8)	EFL = 125 mm, Clear aperture-30 mm	Edmund #47-717
Camera	1.85 μm × 1.85 μm, Sensor size of 4000 × 3000 pixels	iDS, model: U3-3890CP-M-GL Rev.2.2

is obtained, enabling the reconstruction of the sample's three-dimensional surface morphology.

10.3 Setup of Developed CC-DIP

10.3.1 Light Source

The choice of light source for the chromatic confocal system depends on the wavelength range of the chromatic objective. In this research, the chromatic objective operates within a wavelength range of 450–650 nm. Therefore, the light source must provide continuous and uniformly distributed light intensity within its wavelength range. Additionally, for a confocal system, the light source needs to provide sufficient intensity to improve the system's signal-to-noise ratio. This is important to ensure that the image sensor can detect the light intensity signal, even when the sample surface has low reflectivity during actual measurements.

Currently, the industry commonly uses high-intensity broadband light sources such as Xenon lamps and LEDs. LEDs generate broad-spectrum light by exciting fluorescent molecules, typically involving the mixing of blue and yellow light. On the other hand, Xenon lamps generate broadband light directly through an electric arc. In Xenon lamps, the lamp bulb is filled with high-pressure Xenon gas and contains two electrodes. When a high-voltage pulse passes through the electrodes, it ionizes the gas, generating an electric arc. This discharge process excites numerous atoms to higher energy levels, which subsequently release energy as light radiation. This results in a light source with a continuous spectrum. Therefore, Xenon lamps provide more uniform light intensity across a broad wavelength range compared to LEDs, as shown in Fig. 10.2. Taking this advantage into account, a Xenon lamp, combined with a linear fiber optic cable, is chosen as the light source in the proposed system. Figures 10.3 and 10.4 show the adopted light guide cable and Xenon lamp for actual measurements.

The design of the proposed system centers around the key property of the chromatic objective: its dispersion range. Ensuring optimal and stable system performance requires the chosen light source to not only provide sufficient intensity within this wavelength range but also exhibit a continuous and uniformly distributed spectral energy. The design for the illumination module follows a similar approach. To

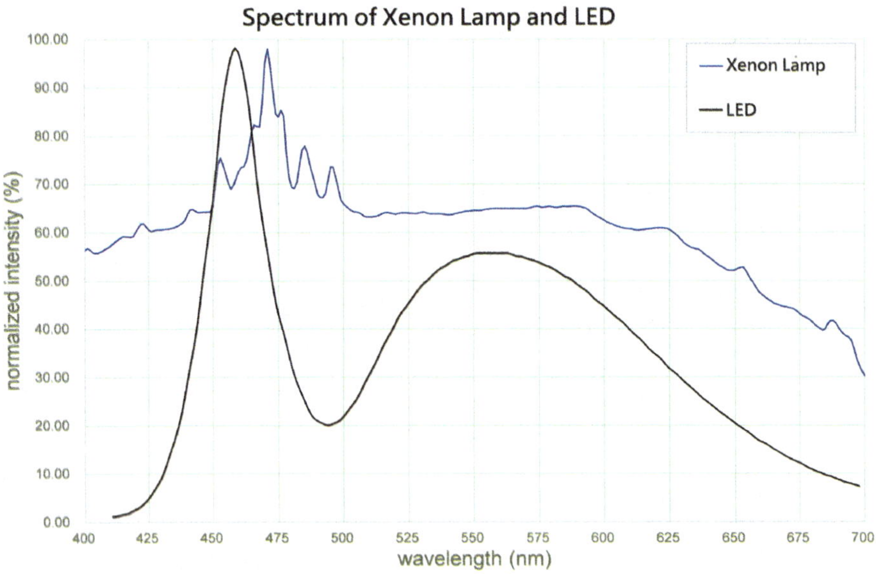

Fig. 10.2 Spectra of Xenon and LED lamp

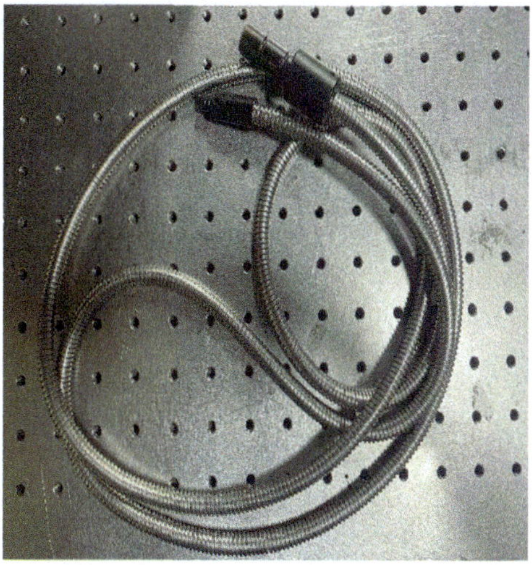

Fig. 10.3 Xenon light guide cable, manufactured by Moritex (model MSG3)

Fig. 10.4 Xenon lamp light source, manufactured by Hamamatsu (model: L9588-03)

fully leverage the capabilities of the dispersion range, the main design specifications adhere to the following conditions:

- Wavelength Range: The illumination module and light source of the system should provide continuous and uniform light intensity across the 450–650 nm range, covering the dispersion range of the chromatic objective.
- Illumination Intensity: To optimize the efficiency and signal-to-noise ratio of the chromatic confocal system, the light source in the illumination module must

provide sufficient intensity. This ensures reliable detection by the image sensor even under low reflectivity conditions of the sample surface.

- Illumination Uniformity: The illumination module should deliver light with uniform intensity distribution to ensure consistent illumination conditions for the objective.
- System Stability: Design considerations for the illumination module should prioritize system stability, including the stability of the light source, thermal management, and noise control. This approach ensures long-term operation reliability and consistent performance.

These design specifications are crucial to ensuring that the illumination module meets the requirements of the chromatic confocal system, thereby enabling high-quality imaging capabilities.

10.3.2 Design and Simulation of Lighting Module

The illumination module was simulated in Zemax to achieve optimal light efficiency in the illumination space. Figure 10.5 shows the Zemax setup, with explanations of the abbreviations provided in Table 10.1. The light guide of the Xenon lamp has an NA of 0.58, necessitating that the collecting lens to have an NA equal to or greater than 0.58. For this purpose, the aspheric lens (Edmund#65-996) with an NA of 0.67 is used. Due to the small diameter of the collecting lens, a diverging lens (ACN254-040-A) was used to expand the beam size. Subsequently, a collimating lens (DLB-40-150PM) and a focusing lens (Edmund#32-925, NA 0.11) were used to focus the light into the illumination space.

A pinhole size of 30 μm was utilized in the illumination space. Both the NA of the illumination module and the NA of the confocal module should be the same. To achieve this, a module was designed for the illumination-object space, as shown in Fig. 10.6. A collimating lens (Edmund #32-923) with a diameter of 40 mm and an effective focal length of 160 mm was employed. The collimated light is then reflected by a nonpolarized beam splitter (BS031) with a size of 2″, providing a 50:50 beam splitting power. The beam size of the collimated light was optimized to completely fill the entrance pupil of the chromatic objective at all wavelengths.

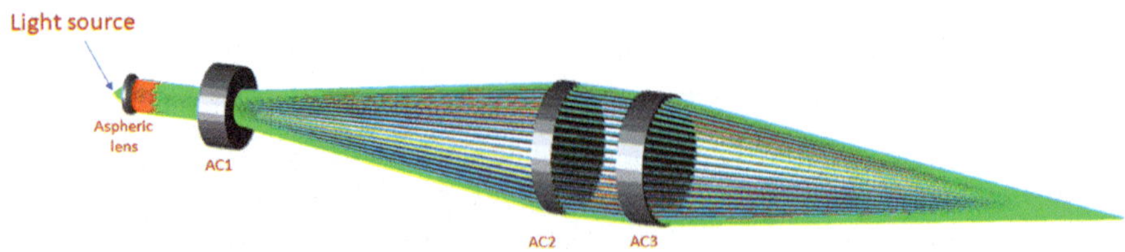

Fig. 10.5 Ray-tracing diagram of the illumination module

Fig. 10.6 Ray-tracing diagram of the illumination-object space module

The chromatic objective focuses broadband light onto the sample. Due to the chromatic nature of the objective, different wavelengths will focus on different positions from the exit pupil, resulting in varying numerical apertures (NA) for different wavelengths.

10.3.3 Illumination and Detection Pinhole

The pinhole size plays a crucial role in the signal-to-noise performance of the chromatic confocal system. A smaller pinhole size theoretically enhances axial resolution. However, smaller sizes also reduce the amount of transmitted light, thereby lowering the signal-to-noise ratio. Therefore, selecting the pinhole size is crucial to achieving the optimal resolution performance.

In the proposed chromatic confocal system, the illumination pinhole size is set to 30 μm. This size is determined based on the magnification of the optical system, as detailed in Chap. 9. The optimized size for the detection pinhole is calculated to be 50 μm, obtained using Eq. (9.4).

10.3.4 Chromatic Confocal Objective

The chromatic objective used in the system, manufactured by PML Lab in NTU, is infinity-corrected and features an NA of 0.72, as shown in Fig. 10.7. The infinity-corrected chromatic objective with a high NA is selected based on the following considerations:

- Enhanced Image Quality: The infinity-corrected design effectively corrects for both chromatic and spherical aberrations, ensuring superior image quality and resolution. This objective delivers sharper and more precise images, minimizing distortions and aberrations across the entire field of view. This attribute is particularly valuable in achieving accurate and high-resolution imaging.
- Increased Light Collection Efficiency: A higher NA enables the collection of a larger amount of light emitted or reflected from the specimen. This increased

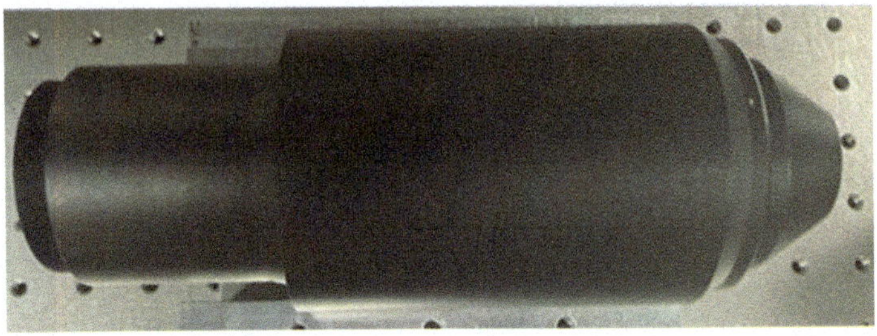

Fig. 10.7 Chromatic confocal objective

light collection efficiency enhances the signal-to-noise ratio, resulting in brighter and more detailed images. By capturing more light, the objective enables the visualization of weakly emitting or low-contrast samples that may be challenging to observe using objectives with low NA.

- Versatility: Infinity-corrected objectives exhibit remarkable versatility and compatibility with various optical components and techniques. Their seamless integration into complex optical systems allows for the utilization of different imaging methods, such as fluorescence microscopy, confocal microscopy, and phase contrast microscopy. This versatility broadens the range of applications and techniques that can be employed.
- Higher Sensitivity and Improved Quantification: The high NA objective enhances the sensitivity of the imaging system, enabling it to detect subtle changes in the specimen more effectively. This increased sensitivity is particularly advantageous in quantitative analysis, facilitating more accurate and reliable measurements of specimen properties.
- Alignment Convenience: The infinity-corrected design also simplifies the alignment and adjustment of optical components. Alignment procedures become more straightforward, aiding in achieving optimal imaging performance and system stability during prolonged experiments.
- Compatibility with Tube Lens Systems: The infinity-corrected chromatic objective is designed to seamlessly integrate with standard tube lens systems commonly employed in microscopy setups. This compatibility ensures that the objective can be readily incorporated into existing microscope configurations without extensive modifications or adjustments.

The Zemax ray-tracing diagram of the chromatic objective is presented in Fig. 10.8. The key optical specifications of the chromatic objective are its dispersion range and the linearity of the dispersion curve, which directly affect the consistency of the depth resolution of the chromatic confocal system across different depth planes.

The dispersion range affects the axial measurement range of the system, with a larger dispersion range enabling a greater depth measurement range. The linearity of the dispersion curve influences the system's stability, as an ideal linear dispersion curve ensures consistent axial resolution at different depths.

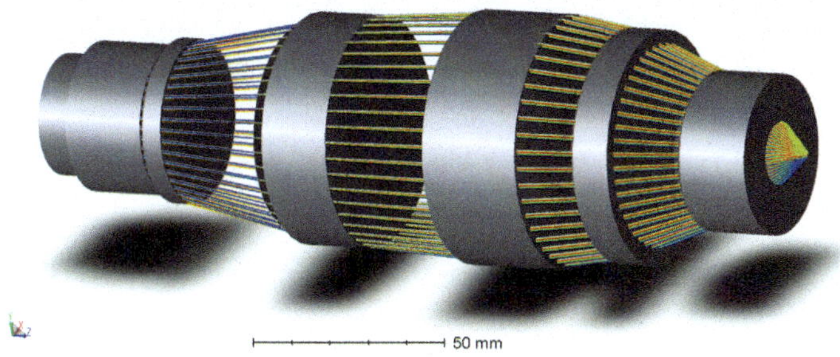

Fig. 10.8 Ray-tracing diagram of the chromatic objective

Figure 10.9 shows the dispersion curve of the infinity-corrected chromatic objective, covering a wavelength range of 450–650 nm with an approximate dispersion distance of 300 μm. Other detailed optical specifications are listed in Table 10.2.

To verify the optical lateral resolution of the infinity-corrected chromatic objective across its entire dispersion range, it is essential to evaluate both the point spread function and the modulation transfer function at different depth planes. Achieving a point spread function and modulation transfer function approaching the diffraction limit indicates good imaging quality. Figures 10.10, 10.11, 10.12, 10.13 and 10.14 show the spot diagram and modulation transfer functions for different depth planes corresponding to wavelengths of 450 nm, 500 nm, 550 nm, 600 nm, and

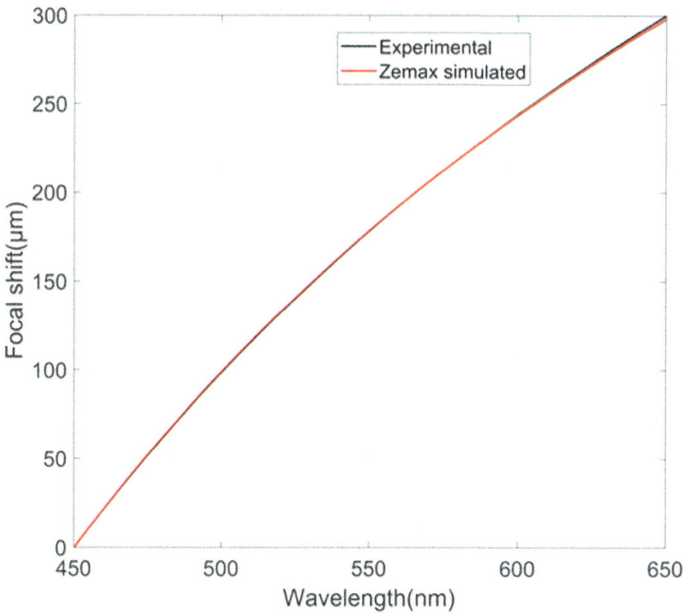

Fig. 10.9 Depth response curve

Table 10.2 Specification of chromatic objective

Working distance (mm)	8.2 (From the last surface to the image)
Field of view (mm)	Ø 1.7
Wavelength range (nm)	450–650
Longitudinal dispersion range	300 μm
Working temperature (°C)	25–35

650 nm, respectively. Based on the data shown in these figures, the infinity-corrected chromatic objective exhibits good optical lateral resolution within its dispersion range.

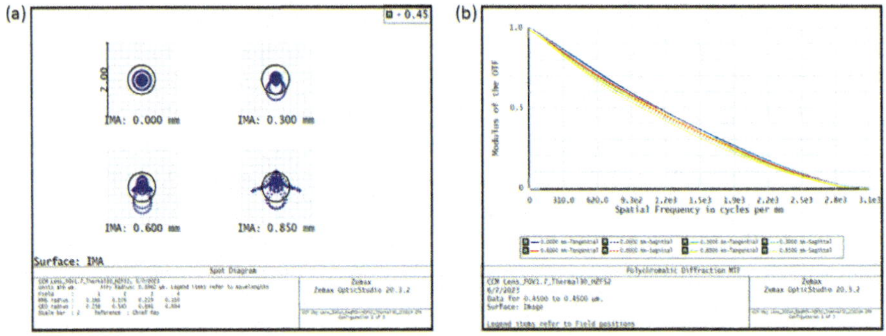

Fig. 10.10 Simulation results: **a** Spot diagram at 0.45 μm; **b** MTF at 0.45 μm

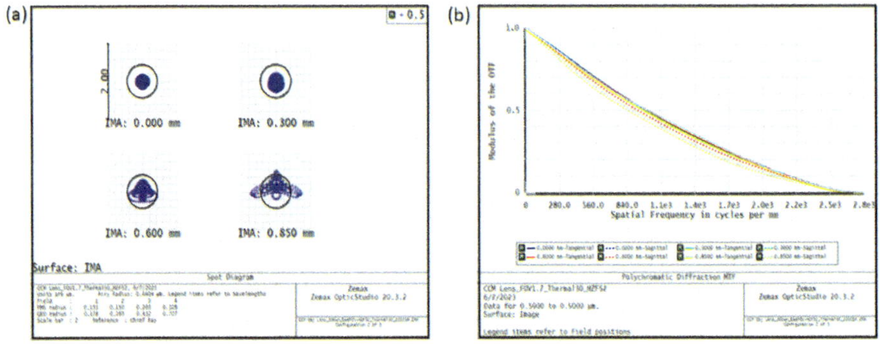

Fig. 10.11 Simulation results: **a** Spot diagram at 0.50 μm; **b** MTF at 0.50 μm

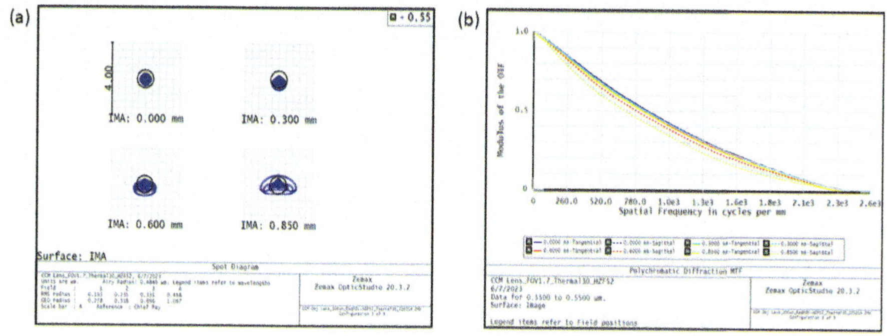

Fig. 10.12 Simulation results: **a** Spot diagram at 0.55 μm; **b** MTF at 0.55 μm

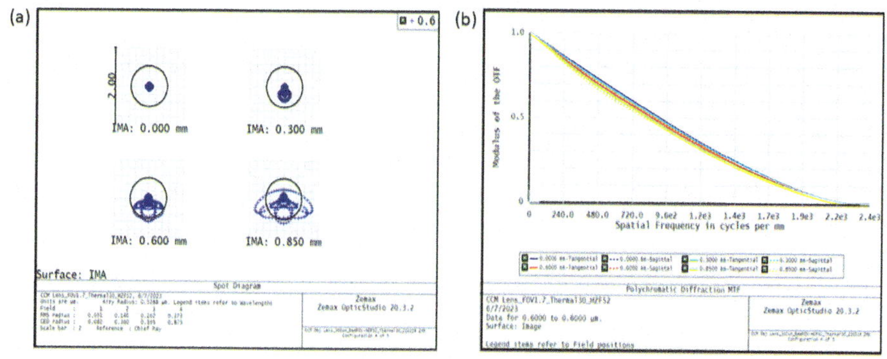

Fig. 10.13 Simulation results: **a** Spot diagram at 0.60 μm; **b** MTF at 0.60 μm

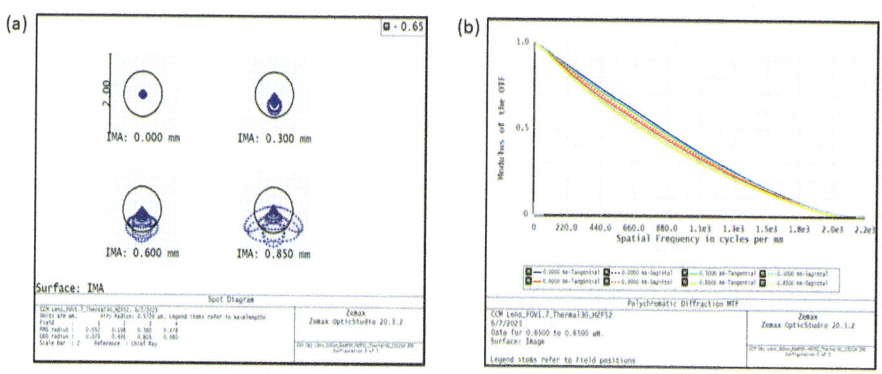

Fig. 10.14 Simulation results: **a** Spot diagram at 0.65 μm; **b** MTF at 0.65 μm

Fig. 10.15 Ray-tracing diagram of the object-detection space

10.3.5 Reflected Path Module

Once the light is reflected from the sample, it passes through the chromatic objective. A tube lens is required to focus the light onto the detector pinhole. The ray-tracing diagram illustrating the reflected path is shown in Fig. 10.15.

The tube lens consists of two achromatic lenses (ACT508-400-A-ML and ACT508-750-A-ML). These lenses have a diameter of 50.8 mm and focal lengths of 400 mm and 750 mm, respectively. The achromatic lens ACT508-400-A-ML operates within the 400–700 nm wavelength range, and the achromatic lens ACT508-750-A-ML has a coating optimized for the same wavelength range.

10.3.6 Sensing Module

In this system, spectroscopic detection is employed to obtain the depth information. The instrument used to receive the spectrum is an imaging spectrometer, which disperses light into its spectral components along a single axis.

The imaging spectrometer adheres to the lens-grating-lens reflective type configuration, as shown in Fig. 10.16. As shown in the figure, light enters the spectrometer through an entrance pinhole. It then passes through a collimating lens, which converts the diverging light from the entrance pinhole into a parallel beam of light. Next, the collimated light encounters a diffraction grating, which diffracts light into its constituent wavelengths, separated based on their diffraction angles. After dispersion, a focusing lens focuses the diffracted light onto a detector. The detector then converts the optical signal into an electrical signal, which is further processed and analyzed by the spectrometer's electronics, creating a spectral profile or spectrum.

The design of the imaging spectrometer adheres to the following criteria:

- **Pinhole**

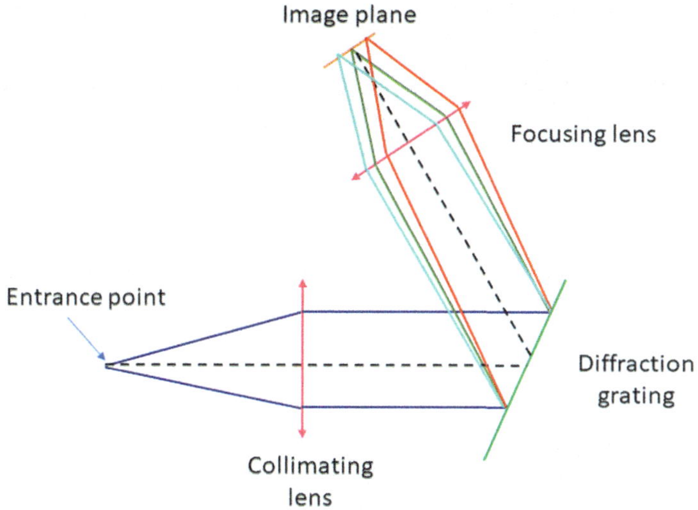

Fig. 10.16 Schematic diagram of the lens-grating-lens type spectrometer

The pinhole used at the entrance of the spectrometer is identical to the detection pinhole. The NA of light focusing on the detector plane should match that entering the spectrometer, which was designed to have an NA of 0.05.

- **Achromatic Lenses**

The inclusion of lenses in a reflective lens-grating-lens design spectrometer offers significant advantages that contribute to its overall performance and functionality. These advantages primarily revolve around light collection, collimation, beam control, and stability. For instance, the lenses in the spectrometer play a critical role in efficiently collecting and focusing the incoming light into the entrance slit, ensuring optimal light-gathering capabilities. Moreover, the lenses help control and stabilize the light beam throughout the optical path, minimizing aberrations, scattering, and unwanted deviations. Overall, the flexibility provided by the lenses enables tailoring the spectrometer's performance to their specific measurement requirements, resulting in desired sensitivity, resolution, and spectral range.

The proposed spectrometer incorporates two achromatic lenses (Edmund #47-719 and Edmund #47-717), each serving different functionalities in the system. The first lens, (Edmund #47-719), collimates the light with a diameter of 30 mm and a focal length of 200 mm. The second achromatic lens, (Edmund #47-717), focuses the light diffracted from the grating onto the camera sensor, which has a larger 300 mm diameter and a 125 mm focal length. These carefully selected achromatic lenses facilitate accurate measurements and ensure optimal performance within the desired spectral range.

- **Diffraction Grating**

In the spectrometer, a reflective diffraction grating disperses light into its constituent wavelengths. It consists of a series of closely spaced parallel grooves on a reflective surface. These grooves act as a diffraction grating, causing light to be diffracted at specific angles determined by the spacing between the grooves and the wavelength of the incident light. For the proposed spectrometer, a diffractive grating of size 25 mm * 25 mm is used (Edmund #64-402), which features 150 grooves/mm.

- **Camera Sensors**

A charge-coupled device (CCD) is employed in the spectrometer to detect the dispersed light from the sample or source. The CCD is a semiconductor device that converts photons into electrical signals, enabling measurement and analysis of the light intensity across different wavelengths. The working principle of a CCD involves the interaction between incident light and the silicon-based sensor array. When photons strike the CCD's surface, they generate photoelectrons through the photoelectric effect. These photoelectrons are then accumulated and stored in potential wells within the CCD's structure. The CCD consists of an array of pixels, each acting as an individual light-sensing element. Each pixel collects and quantifies the number of photoelectrons generated by the incident light at a specific location. The accumulated charge in each pixel is proportional to the intensity of light falling on it. To read out the accumulated charges, a clocking mechanism shifts the charges from pixel to pixel within the CCD array, eventually transferring them to an output amplifier. The charges are then converted into voltage signals, amplified, and digitized for further processing and analysis.

To maintain good resolution while achieving high-speed measurement capabilities, the proposed system employs a monochrome CMOS sensor commonly available in the market. The sensor, manufactured by iDS, is the U3-3890CP-M-GL Rev.2.2 model with a pixel size of 1.85 μm × 1.85 μm, a sensor size of 4000 × 3000 pixels, and a maximum frame rate of 33.0 fps.

- **Optical Performance of Sensing Module**

The ray-tracing simulation diagram of the sensing module is depicted in Fig. 10.17, showing the arrangement of the applied components. Note that the blaze angle of the blazed grating is 2.13°, corresponding to a coverage of 2087 pixels in the working range when a sensor with a pixel size of 1.85 μm is used.

10.4 Zemax Integration of the Developed Confocal System

The confocal system's integration was performed using Zemax nonsequential mode to accurately model all its components. To verify its performance, the system was tested at five different BFL (Back Focal Length) depths in the object space: 10.032 μm, 10.131 μm, 10.211 μm, 10.275 μm, and 10.330 μm, as shown in Fig. 10.18a–e, respectively. These depths are measured from the object's surface to the last surface of the objective lens, each corresponding to the wavelengths of

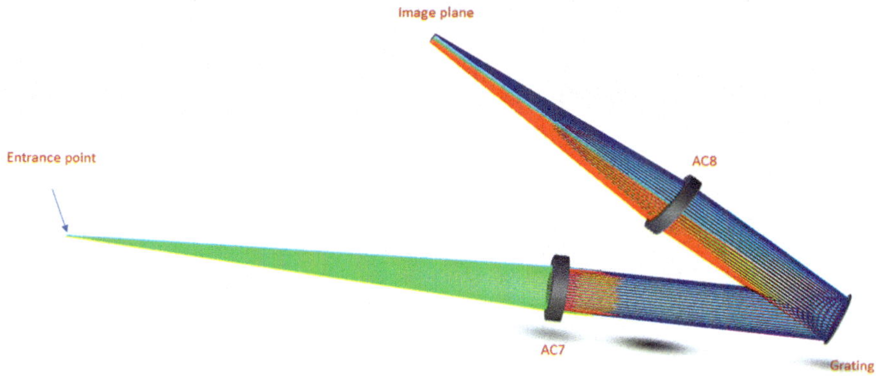

Fig. 10.17 Ray-tracing diagram of the spectrometer

450 nm, 500 nm, 550 nm, 600 nm, and 650 nm, respectively. The sensing module maps different wavelengths for each depth, as depicted in the figure. The simulation results from Zemax demonstrate that the confocal system is capable of accurately measuring depth. The actual setup of the confocal optical system on the optical bench is given in Fig. 10.19.

10.5 Commercial Spectrometer

The spectrum obtained from the designed imaging spectrometer was compared with a calibrated commercial spectrometer shown in Fig. 10.20. Its specifications are given in Table 10.3 [1].

The optical fiber used with the spectrometer is the OCEAN OPTICS 727-733-2447, featuring a fiber core diameter of 50 μm and a wavelength range of 300–1100 nm. The photo and specification of the fiber are provided in Fig. 10.21 and Table 10.4 [2].

10.6 Scanning Stage

The scanning stage is employed to scan the sample in the object space, providing movement along the *X*-axis and *Y*-axis. It is equipped with a stepper motor, enabling precise movement control through discrete steps. The stage's movement is managed via a graphical user interface (GUI), as detailed in the next section. Figure 10.22 shows the scanning stage utilized in the experiment.

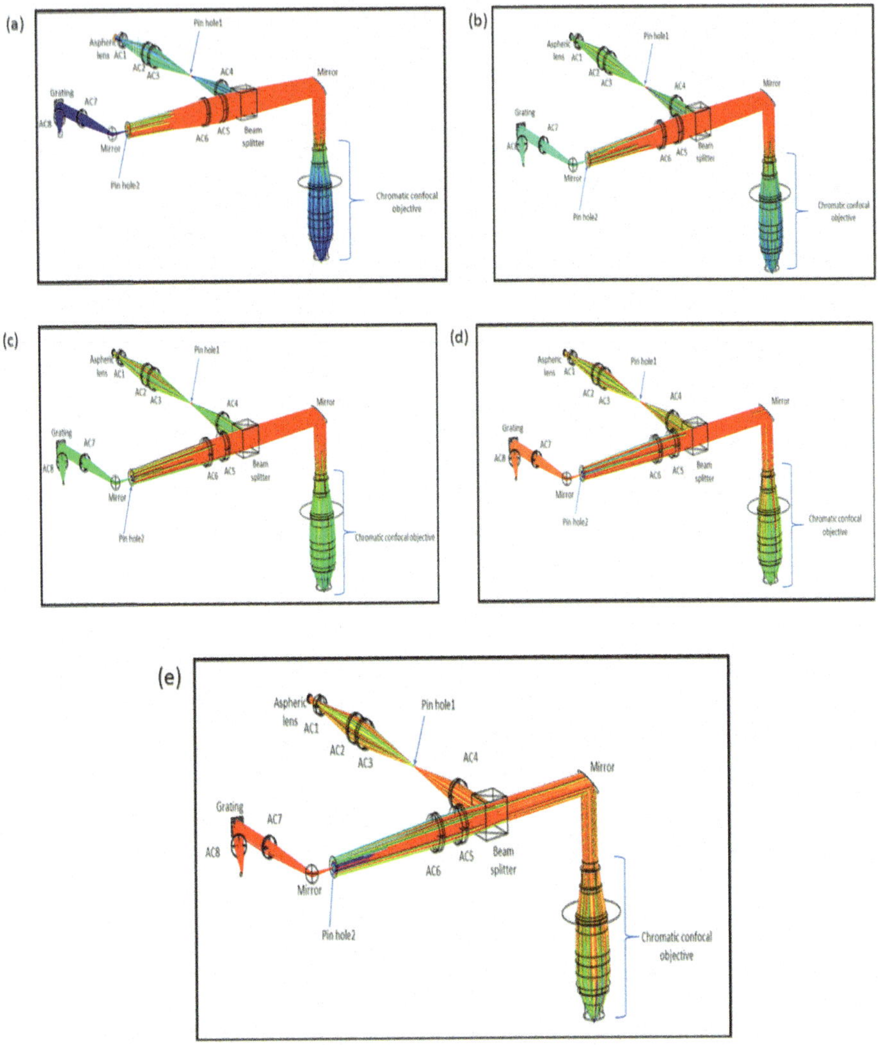

Fig. 10.18 Zemax simulation diagram confocal setup **a** BFL = 10.032 μm; **b** BFL = 10.131 μm; **c** BFL = 10.211 μm; **d** BFL = 10.275 μm; **e** BFL = 10.330 μm

10.7 Control Software and GUI

To synchronize and control the hardware devices for automated measurements, the C++ programming language and the QT framework were used to develop a GUI that integrates the system's measurement functions, as shown in Fig. 10.23. Through this interface, users can connect the camera, the piezoelectric stage, and the step motorized displacement platform. The software allows functions such as camera

Fig. 10.19 Optical setup of confocal system

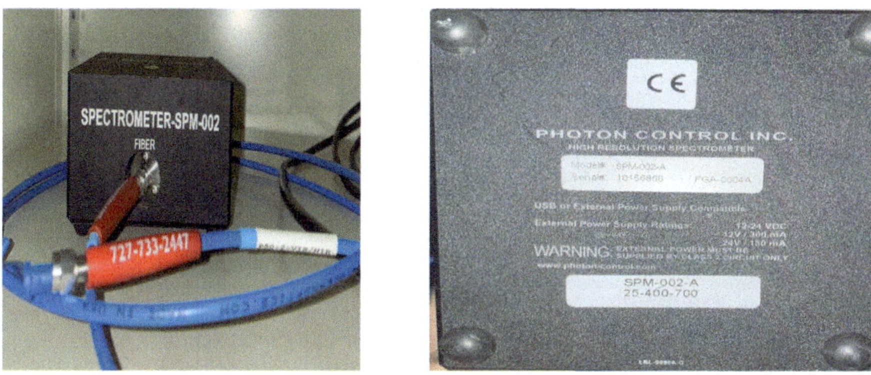

Fig. 10.20 Photo of the commercial spectrometer (model name: SPM-002-A)

Table 10.3 Specification of photon control spectrometer SPM-OO2-XH [1]

Technical specification of SPM-002-A	
Spectral range (nm)	400–700
Resolution (center/outer)	0.5/0.6
Standard slit (μm)	25
Detector type	CCD
Pixels	3648
Focal length (mm)	70
Exposure time range (μs)	16–65

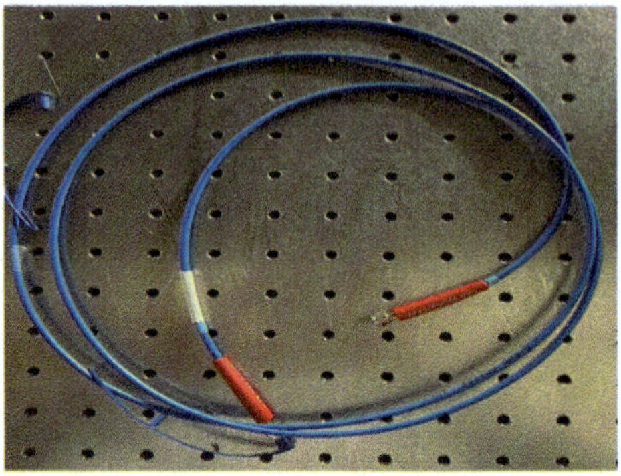

Fig. 10.21 Photo of the spectrometer fiber

Table 10.4 Specification of ocean optics fiber

Wavelength range (nm)	300–1100
Fiber core size (μm)	50
Length (m)	2
Jacket	PVDF zip tube
Buffer material	Polyamide
Cladding OD[a] (μm)	35
Cladding OD tolerance (μm)	± 0.5
Connector	SMA905
Fiber core size tolerance (μm)	± 5
Long term bend radius (cm)	4
Maximum OD (μm)	155
Primary buffer OD (μm)	17
Primary buffer OD tolerance (μm)	± 5

[a] OD—Outer diameter

parameter adjustment (e.g., exposure time and region of interest), real-time image display, and implementation of depth correction algorithms.

10.8 Comparison of Commercial and Experimental Spectrum

To validate the experimental spectrum measured by the designed imaging spectrometer, the results are compared with those obtained from a calibrated commercial spectrometer, as shown in Fig. 10.24. Table 10.5 shows the comparison of FWHM values at different wavelengths.

Fig. 10.22 Scanning stage used in the experiment

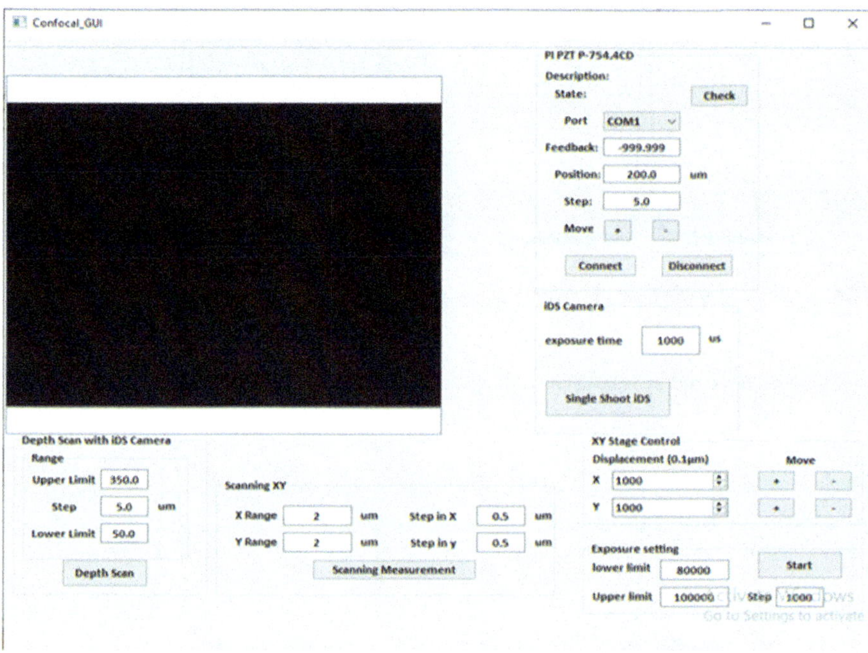

Fig. 10.23 GUI for the confocal system

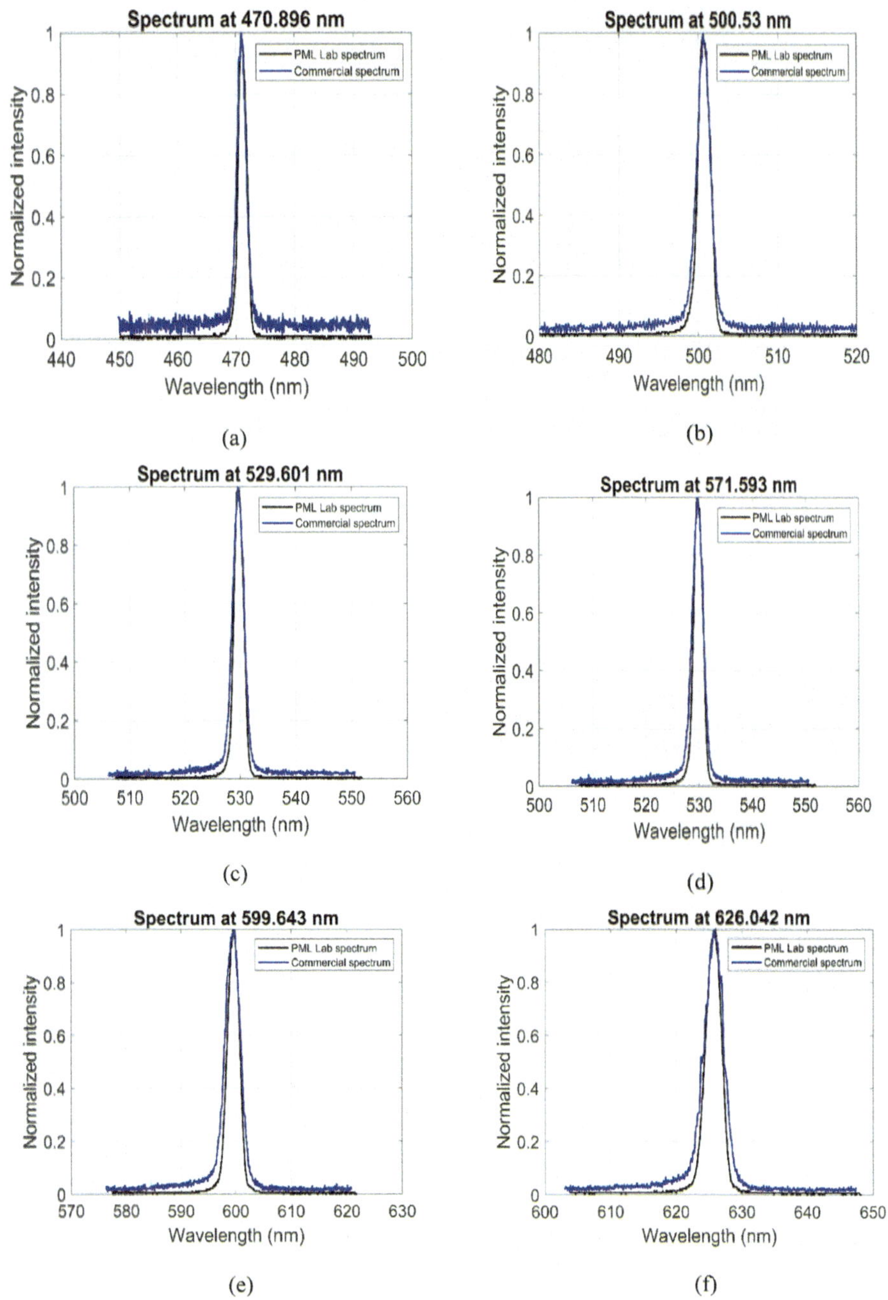

Fig. 10.24 Comparison of the measured spectrum at different wavelengths

Table 10.5 FWHM comparison of the measured spectrum

		FWHM (nm)	Deviation (nm)	% Deviation
Spectrum at 470.896 nm	Commercial	1.86		
	PML Lab	1.72	0.14	7.52
Spectrum at 500.53 nm	Commercial	1.95		
	PML Lab	1.88	0.067	3.5
Spectrum at 529.601 nm	Commercial	2.39		
	PML Lab	2.05	0.34	14.22
Spectrum at 571.593 nm	Commercial	2.60		
	PML Lab	2.41	0.18	7.3
Spectrum at 599.643 nm	Commercial	2.92		
	PML Lab	2.63	0.28	9.93
Spectrum at 626.042 nm	Commercial	3.05		
	PML Lab	2.84	0.20	6.88

10.9 Algorithm and System Calibration

10.9.1 Peak Detection

Since the pixels on the photodetector provide discrete information, the spectral signal cannot accurately reflect the peak positions within these discrete sampling points. Therefore, data fitting is required to fit the data to a specific function, ensuring the correct spatial positions of the spectral peaks are obtained.

The spectral image, as shown in Fig. 10.25, has the spatial coordinate on the horizontal axis and the spectral coordinate on the vertical axis. Different vertical positions correlate to different wavelengths and corresponding depths, while the grayscale values represent light intensity. To obtain the depth information for each spatial position, a profile along the vertical axis of the spectral image can be plotted. This profile includes pixel indices representing the wavelengths and their corresponding depths obtained from the depth response curve, as detailed in the next section.

Before performing the peak detection algorithm, a signal filter is applied to eliminate noise, and an appropriate threshold is set to reduce the impact of data points with low signal-to-noise ratios on peak detection. Finally, a curve-fitting algorithm is used to determine peak positions.

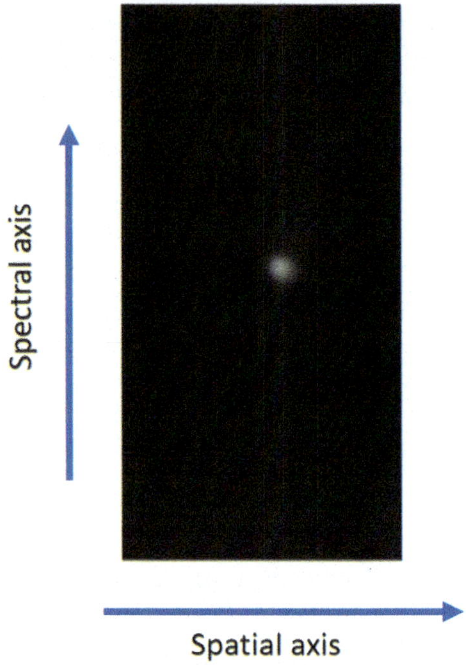

Fig. 10.25 Spectrum obtained on the camera sensor

10.9.2 Depth Calibration

Establishing a depth calibration curve is a crucial step in the operation of a chromatic confocal system. The depth calibration curve serves as the system's database, converting wavelengths measured by the spectrometer into spatial depth information. Therefore, the quality of the depth calibration curve significantly affects the interpretation of depth positions during measurements. Ideally, the depth calibration curve should follow a strictly increasing or decreasing monotonic function. Failure to exhibit such a distribution could lead to ambiguity, where a single wavelength corresponds to multiple depth values.

The method to establish the depth calibration curve involves fixing a flat mirror onto a high-precision piezoelectric translator (PZT). The position feedback from the PZT ensures precise and reliable depth positioning with nanometer-level precision and resolution.

Performing the calibration process is crucial to ensure accurate and reliable measurements. Calibration plays a crucial role in reducing inherent system errors that may arise during the manufacturing and assembly of the optical measurement system. With proper system calibration, various factors that may introduce discrepancies between the actual system performance and the values predicted by theoretical models, including variations in system components, wavelength-dependent aberrations, nonuniform system spectral response, etc., are considered and compensated for, leading to more precise and consistent measurements.

The flowchart for the calibration process is shown in Fig. 10.26. The calibration begins with the arrangement of a calibration surface, which serves as a reference for accurate measurements. This surface should be carefully prepared to ensure stability.

Next, the piezo is positioned at the farthest side of the measurement range to ensure complete coverage during calibration. The piezo is then incrementally moved in 5 μm steps toward the edge of the measurement range. At each step, 30 spectral images are captured and averaged to reduce noise and improve measurement accuracy. Once the spectral images are obtained, a peak extraction algorithm is applied to identify peaks corresponding to different wavelengths. The applied algorithm precisely locates peak positions and extracts relevant information from the captured data. Finally, a polynomial fitting method is used to calibrate the system based on the extracted peak information. This fitting process establishes the relationship between measured data and actual values, providing a calibration result that can be used for correcting system-induced errors during subsequent measurements.

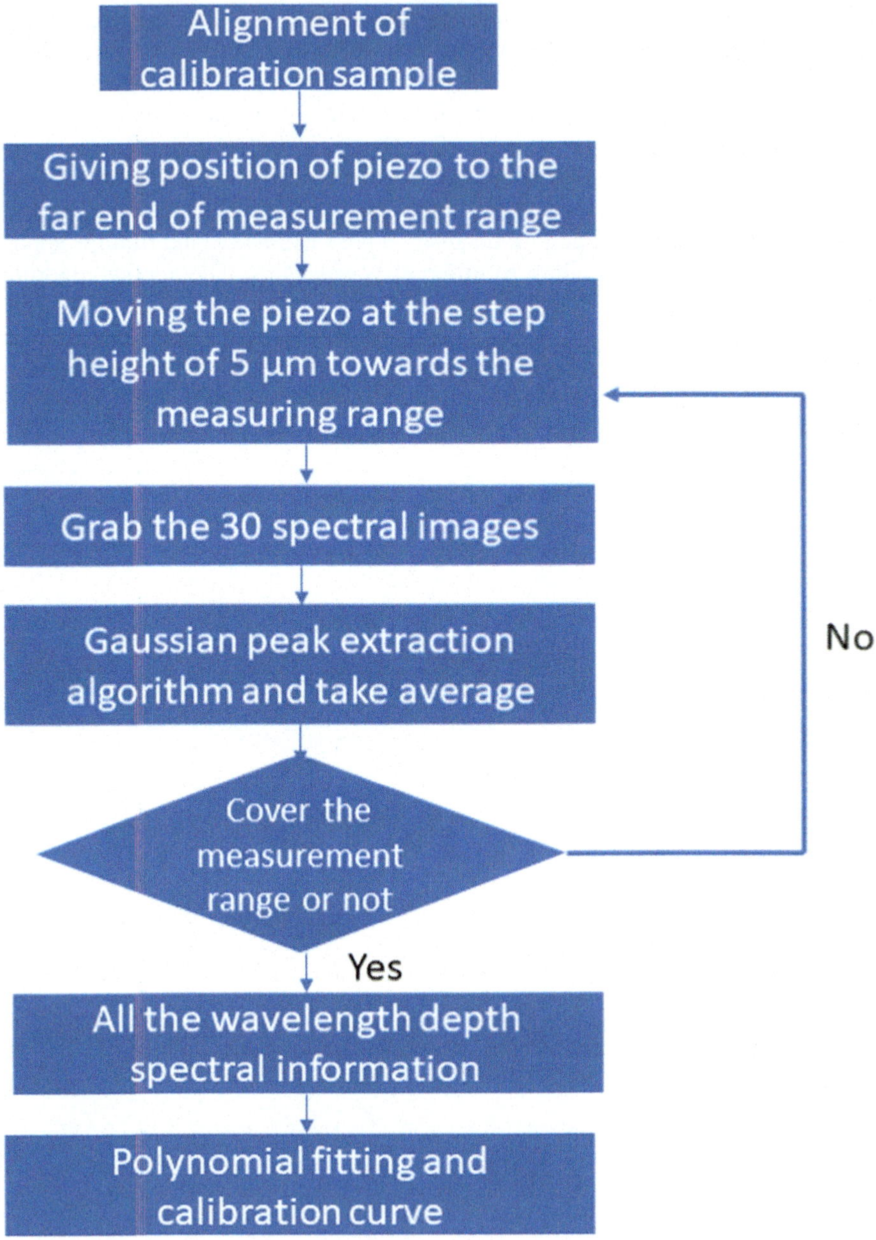

Fig. 10.26 Flowchart of the calibration process

References

1. PhotonControl (2023) Technical specification of spectrometer. SPM-002-XH. https://wiki.hackuarium.ch/images/f/f6/Spm002.pdf. Accessed 2023
2. Optics O (2023) Ocean optics fiber optic cable 2m 727-733-2447 P50-2-UV-VIS For Spectrometer. https://www.oceaninsight.com/products/fibers-and-probes/fibers/patch-cords/p50-2-uv-vis/. Accessed 2023
3. Singh SK (2023) Theory and implementation of microscopic diffractive image profilometry. Master thesis, National Taiwan University

Chapter 11
Applications of Chromatic Confocal Diffractive Image Profilometry (CC-DIP)

11.1 Introduction

In the preceding chapter, the realization of chromatic confocal diffractive image profilometry (CC-DIP) was explained. The primary objective of this chapter is to present the results obtained from experimental analyses to demonstrate how deep learning deconvolution contributes to enhancing accuracy and resolution both axially and laterally.

11.2 Light Intensity

Within the system, various optical components, from the light source spectrum to different optical elements, exhibit distinct spectral behaviors. These discrepancies lead to inconsistent baseline light intensities at varying depths, influencing the measurement accuracy. Moreover, lower light intensities are particularly vulnerable to noise interference, resulting in uneven measurement performance at different depths.

To assess the system's sensitivity to light intensity variations across different wavelengths, a high-reflectivity flat mirror was employed as the measurement sample to minimize the impact of sample reflectance on wavelength changes [1]. The flat mirror was positioned on a piezoelectric displacement platform, and measurements were taken at a single depth with a spacing of 5 μm over a displacement of 300 μm. Thirty measurements were conducted at this depth, and the average grayscale value of the peak on the measurement line was calculated to mitigate random measurement errors.

The validation results, presented in Fig. 11.1, illustrate the average light intensity values at various depth positions, starting from the focal plane at 300 μm for a wavelength of 650 nm. The relative light intensity at each depth was determined

L. Chen et al., *Diffractive Image Microscopy for 3D Imaging*, Springer Tracts in Mechanical Engineering, https://doi.org/10.1007/978-981-97-7782-2_11

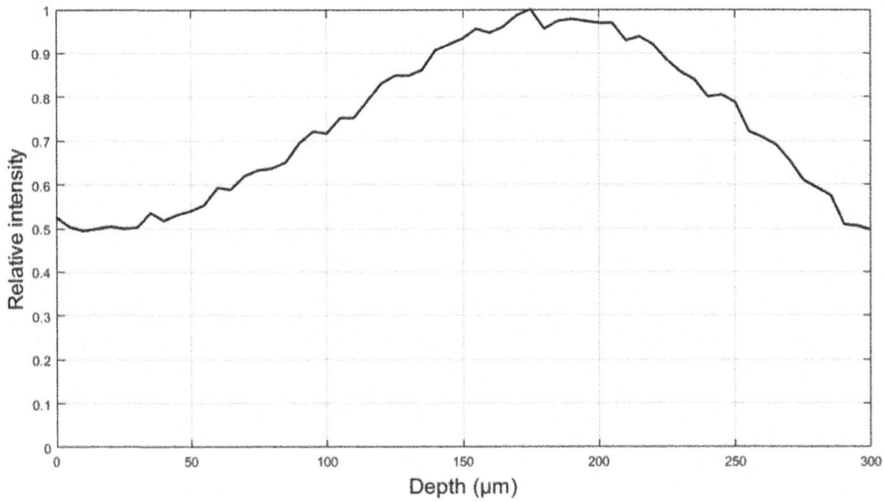

Fig. 11.1 Relative light intensity response at different depths

based on the maximum grayscale value in the data. The validation graph indicates that the average variation in light intensity response across different wavelengths is 0.8. Moreover, 80% of the measured range exhibits light intensity ratios exceeding 0.7, with the maximum disparity in light intensity occurring at the boundaries of the measurement range.

11.3 Tilt Measurement of the Object Space

Following the system calibration process described in Chap. 10, it is crucial to examine any potential tilt in the object space, as it could potentially affect the accuracy of the measurement results [1]. To evaluate tilt, a flat mirror was scanned across a 200 μm × 200 μm area. The resulting surface plot diagram is shown in Fig. 11.2.

11.4 Polynomial Fitting Equation

The depth measurement range of the system primarily depends on the axial dispersion range of the chromatic objective. In this study, the applied chromatic objective has a dispersion range of approximately 300 μm within the 450–650 nm wavelength range. To validate the depth measurement range and the depth calibration curve of the objective, a flat mirror was fixed on the piezoelectric displacement platform, and the platform's position feedback was used for positioning [1]. The platform displacement was set at intervals of 5 μm, covering a total depth displacement of

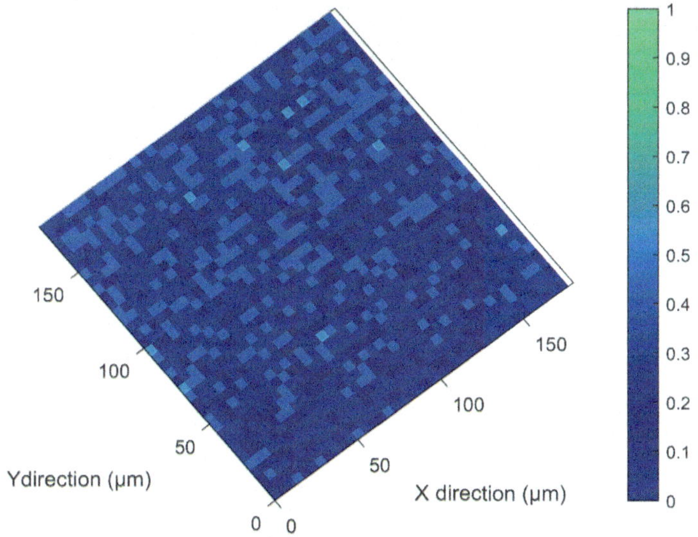

Fig. 11.2 Measurement of the tilt error of a flat surface

300 μm. Initially, polynomial fittings of various orders, ranging from second to ninth order, were applied to the depth response data, as seen in Table 11.1. The sum square error (SSE) was used as a metric to evaluate the goodness of fit for each order. Interestingly, it was observed that the SSE reached a saturation point after the fourth order, suggesting that higher-order polynomial fittings did not significantly improve the performance, as plotted in Fig. 11.3. Consequently, the fourth-order polynomial fitting was selected as the optimal choice.

Note that polynomial fitting is crucial for capturing the underlying trend and relationship between variables. It accommodates complex nonlinearities and higher-order dependencies present in the depth response curve, ensuring an accurate representation of the system's behavior and the extraction of valuable insights from the data. Moreover, polynomial fitting enables extrapolation and interpolation of

Table 11.1 Polynomial fitting order and SSE value

Order	SSE (μm)
2	75.105
3	0.529325
4	0.0984
5	0.0828
6	0.0788
7	0.0772
8	0.0693
9	0.0413

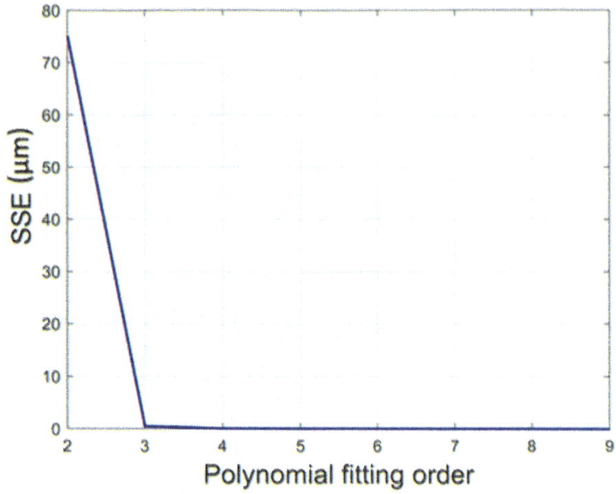

Fig. 11.3 Plot of different orders of polynomial fitting

the depth response curve beyond the measured data points, thereby enhancing the system's depth measurement capabilities. Overall, the fitted polynomial model provides a concise and interpretable representation of the depth response, facilitating system characterization and calibration.

The polynomial fitting of the depth response curve obtained from the chromatic objective is given in Fig. 11.4. The coefficients of polynomial fitting are provided in Table 11.2, and the deviation between the raw data and the polynomial fit is calculated and plotted in Fig. 11.5.

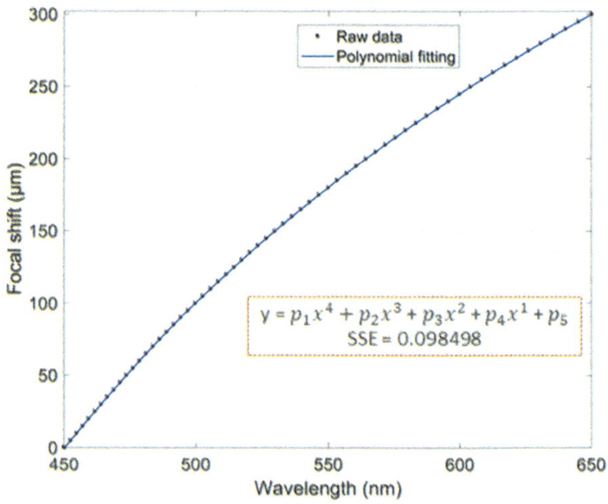

Fig. 11.4 Polynomial fitting of depth response curve

Table 11.2 Coefficients of polynomial fitting

Coefficients of polynomial fitting	Value
p_1	− 0.1235
p_2	1.512
p_3	− 11.1
p_4	88.79
p_5	160.7
The variable (x) is normalized by mean 536.6 and std 58.65	

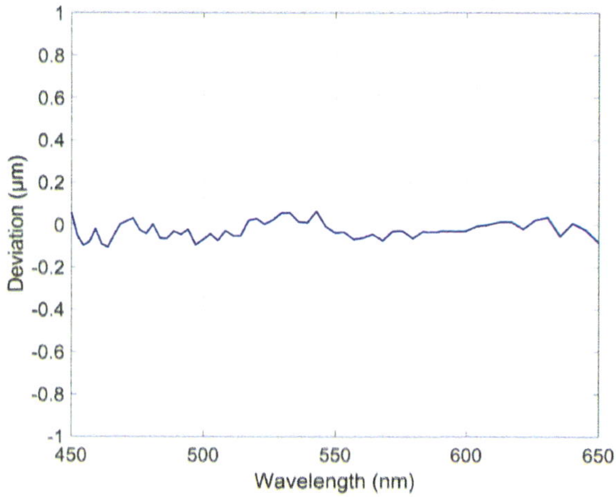

Fig. 11.5 Deviation between the fourth-order polynomial fitting and the raw data

11.5 Sample Measurement

11.5.1 Step Height Measurement

We conducted measurements with our experimental setup to determine the step heights of 1, 5, and 10 μm. Ten sets of measurement data are collected to assess the repeatability and stability of the measured results. Each step height was measured twice: once before deconvolution and once after deconvolution.

The measured height values of the standard step height were compared to data obtained from a white light interferometer installed in the PML Lab to analyze bias in the proposed measurement system [1].

1. 10-μm step height

The stability of the measurement for a 10-μm step height was evaluated through ten repeated measurements. The experimentally obtained values were used to calculate the standard deviation and average value, presented in Table 11.3.

Figure 11.6 shows the cross section of the measured step height, with the fitted data representing the average value of the measured surface.

Table 11.3 Raw data of measured profile for the 10-μm step height

10-μm measured step height data (μm)		
10.37	10.358	10.39
10.403	10.362	10.27
10.378	10.391	10.391
10.411		
Average value (μm)	10.372	
Standard deviation (μm)	0.039	

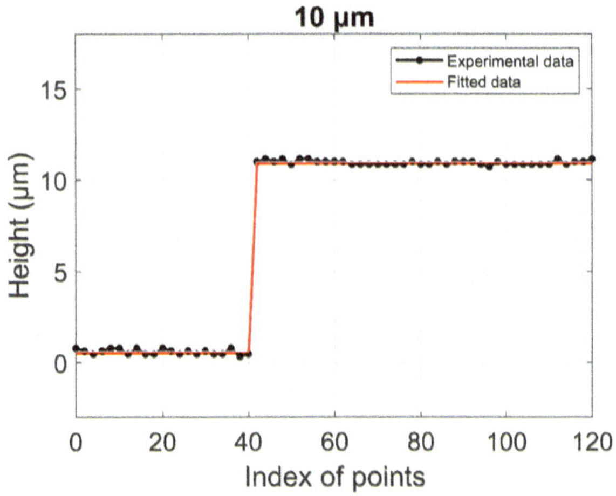

Fig. 11.6 A cross-section plot showcasing the experimentally obtained step height of the 10-μm step height

The experimental results yielded a spectrum that was then subjected to deconvolution. The deconvolution method was applied to obtain results from the ten measurements, where the deconvolved values, standard deviation, and average are provided in Table 11.4. Figure 11.7 illustrates the plot of the 10 μm step height, with another figure showing its cross section. The fitted data represents the average value of the measured surface.

Table 11.4 Deconvolved result data of profile for the 10-μm step height

10-μm deconvolved step height data (μm)		
10.306	10.372	10.354
10.363	10.387	10.447
10.617	10.614	10.599
10.601		
Average value (μm)	10.466	
Standard deviation (μm)	0.126	

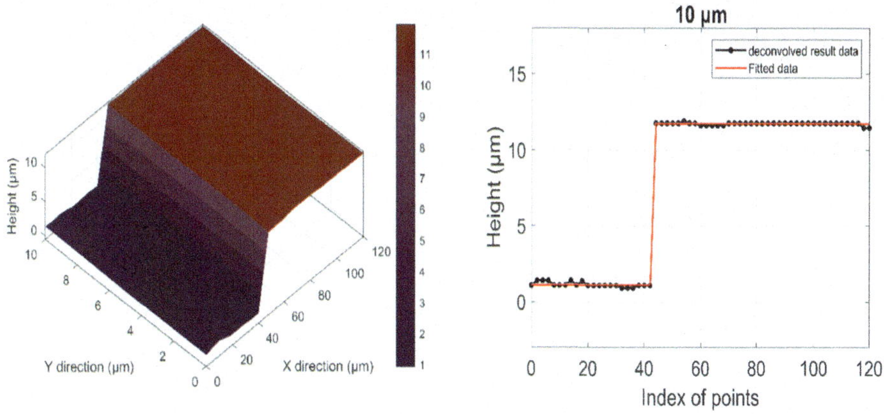

(a) (b)

Fig. 11.7 **a** Three-dimensional plot of the 10-μm step height obtained from deconvolution; **b** cross-section plot of the 10-μm step height obtained from deconvolution

2. 5-μm step height

The measurement stability for a 5-μm step height was evaluated by conducting ten repeated measurements. The experimentally obtained values were used to calculate the standard deviation and average value, presented in Table 11.5. Figure 11.8 displays the cross section of the measured step height, where the fitted data represents the average value of the measured surface.

The experimental results yielded a spectrum that was subsequently deconvolved. The deconvolution method was applied to obtain results for the ten measurements. These values, along with their standard deviation and average, are provided in Table 11.6. Figure 11.9 illustrates the plot of the 5-μm step height, while another figure presents the cross section. The fitted data represents the average value for the measured surface.

3. 1-μm step height

Table 11.5 Raw data of measured profile for the 5-μm step height

5-μm measured step height data (μm)		
5.497	5.502	5.44
5.519	5.544	5.495
5.483	5.481	5.469
5.488		
Average value (μm)	5.4918	
Standard deviation (μm)	0.027	

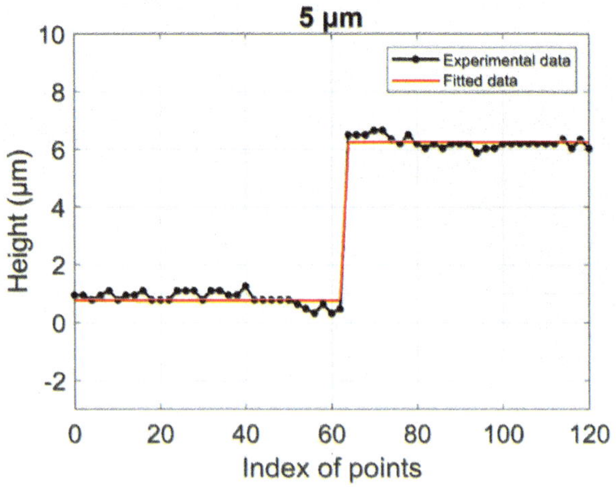

Fig. 11.8 A cross-section plot showcasing the experimentally obtained step height of the 5-μm step height

Table 11.6 Deconvolved result data of profile for the 5-μm step height

5-μm deconvolved step height data (μm)		
5.528	5.535	5.555
5.552	5.558	5.529
5.543	5.516	5.546
5.542		
Average value (μm)	5.540	
Standard deviation (μm)	0.013	

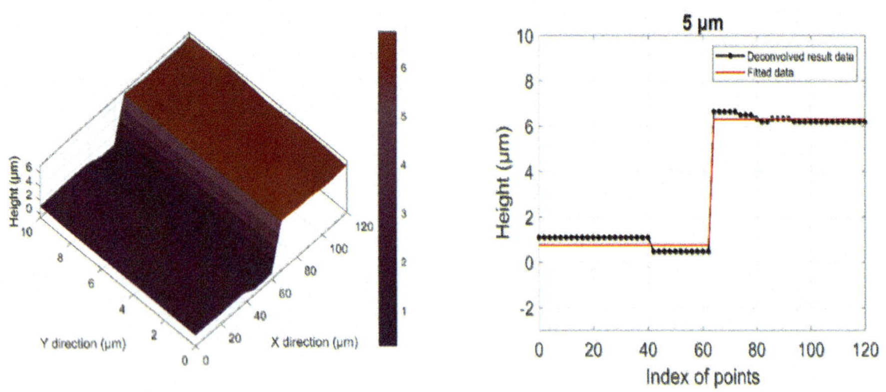

(a) (b)

Fig. 11.9 **a** Three-dimensional plot of 5 μm step height obtained from deconvolution; **b** cross-section plot of the 5 μm step height obtained from deconvolution

Table 11.7 Raw data of measured profile for 1-μm step height

1-μm measured step height data (μm)		
0.737	0.736	0.729
0.713	0.713	0.73
0.728	0.727	0.726
0.713		
Average value (μm)	0.725	
Standard deviation (μm)	0.009	

The measurement stability for a 1-μm step height was evaluated by conducting ten repeated measurements. The experimentally obtained values were used to calculate the standard deviation and average value, presented in Table 11.7. Figure 11.10 shows the cross section of the measured step height. The fitted data represents the average value of the measured surface.

Following similar approaches as the preceding sections, the deconvolved values, along with their standard deviation and average, are provided in Table 11.8. Figure 11.11 illustrates the plot of the 1-μm step height, while another figure presents its cross section. The fitted data represents the average value for the measured surface.

The higher standard deviation observed in the deconvolved results compared to the experimental ones can be attributed to the nature of the deconvolution process. Deep learning Weiner deconvolution, while effective in enhancing the accuracy of the measured step height, introduces additional uncertainties due to the deblurring process. During deconvolution, noise and artifacts present in the original spectrum can be amplified, leading to an increase in the standard deviation of the deconvolved data. On the other hand, the experimental measurements involve fewer computational steps and potential sources of uncertainty. As a result, the measured standard deviation is comparatively lower.

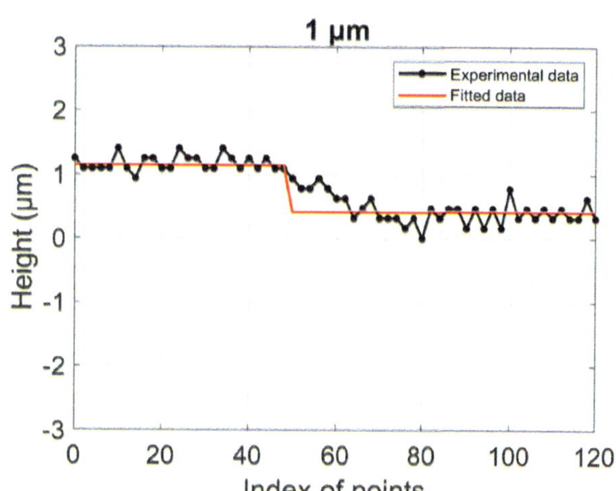

Fig. 11.10 A cross-section plot showcasing the experimentally obtained step height of the 1-μm step height

Table 11.8 Deconvolved result data of profile for 1-μm step height

1-um deconvolved step height data (μm)		
0.946	0.957	0.969
0.975	0.978	1.003
1.011	1	1.008
0.997		
Average value (μm)	0.984	
Standard deviation (μm)	0.022	

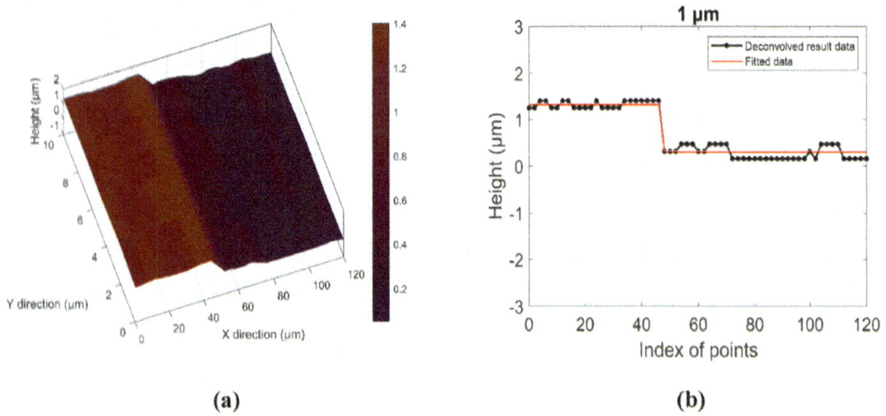

Fig. 11.11 **a** Three-dimensional plot of the 5-μm step height obtained from deconvolution; **b** cross-section plot of the 5-μm step height obtained from deconvolution

It is important to consider that deconvolution is a mathematical technique, and its efficacy is influenced by various factors, including the quality of the original spectrum, the chosen deconvolution algorithm, and the calibration of the system. These factors can collectively contribute to the higher standard deviation in the deconvolved results.

Despite the slightly higher standard deviation, it is crucial to note that the deconvolved results exhibit closer agreement with the actual step height values. The deconvolution process effectively enhances the measurement resolution and accuracy, enabling better identification of peak positions and finer details in the spectrum. Thus, the trade-offs of using deep learning deconvolution should be acknowledged depending on specific applications.

In conclusion, for the measurement of small step heights such as 1 μm, the deconvolution technique demonstrates its effectiveness in accurately detecting the step height edges. The results obtained from deconvolution closely align with the measurements obtained using white light interferometry. This indicates that deconvolution is a reliable method for enhancing the precision and accuracy of step height measurements, particularly for smaller step heights.

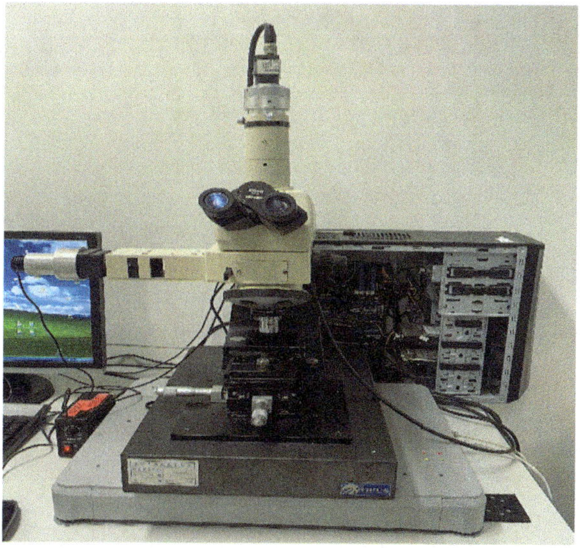

Fig. 11.12 White light interferometry at PML Lab

11.5.2 WLI for Standard Measurement of Step Height

The height values of the standard step heights measured in Sect. 11.5.1 were compared with the data obtained from a WLI located in the PML Lab, as depicted in Fig. 11.12.

The white light interferometer utilized a 20× Mirau objective lens with NA 0.4, enabling highly precise step height measurements. Hence, the step height depths measured by this commercial instrument serve as benchmark values for assessing the accuracy deviation (bias) of the proposed system. The profile cross sections obtained from WLI are given in Fig. 11.13. Detailed comparative data can be found in Table 11.9.

11.5.3 Grating Structure

A silicon wafer sample was measured to analyze its grating-like structure. The grating structure is characterized by its periodicity and sharp edges. The dimensions of the grating measured from the proposed system were compared with the data obtained from an AFM. The values measured by the AFM serve as a reference for analyzing bias in the proposed measurement system [1].

The grating structure under investigation has a nominal dimension of W8/S8, where "W" represents the width and "S" represents the spacing. The nominal depth of the structure is 5 μm. Due to its confined nature, it is essential to ensure that the

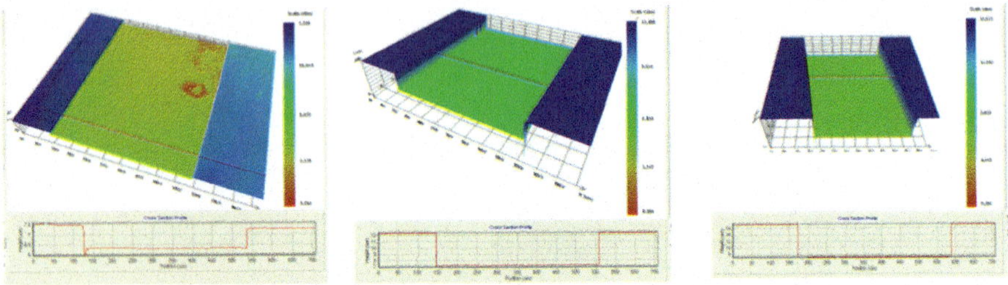

(a) (b) (c)

Fig. 11.13 WLI measured results: **a** Three-dimensional view of 1-μm, **b** 5-μm, and **c** 10-μm step height obtained from WLI

Table 11.9 Measurement results of step height blocks at different heights

Step height	10 μm	5 μm	1 μm
WLI (μm)	10.282	5.386	1.08
Deconvolved result (μm)	10.466	5.540	0.984
Bias (μm)	0.184	0.154	0.096
Standard deviation (μm)	0.126	0.013	0.022

incident light from the focused objective reaches the bottom surface of the structure and reflects back to the objective during chromatic confocal microscopy measurements. This enables the reconstruction of a three-dimensional profile of the measured sample.

When sample dimensions shrink to the micrometer scale or smaller, only a fraction of the incident light reaches the bottom surface. Moreover, the reflected signal from the bottom surface is further attenuated by the side walls of the sample [2]. As a result, the signal-to-noise ratio decreases significantly, making it insufficient for accurate reconstruction of the sample structure.

Using a high NA objective lens exacerbates this issue. The high NA objective focuses light with a higher cone angle onto the top surface of the sample, leading to a substantial amount of light energy loss from the bottom surface of the structure.

In our measurements, we employ a chromatic objective with an NA of 0.7. While this NA value is considered high for measuring confined structures, it helps to optimize the measurement conditions for improved accuracy and signal quality.

The experimental spectrum obtained from the measurement of the W8/S8 sample is presented in Fig. 11.14. It is evident from the plotted spectrum that the signal intensity has a low signal-to-noise ratio, making it challenging to extract accurate information from the signal. Moreover, traditional deconvolution methods would yield only limited improvements.

Also presented in Fig. 11.14, the resulting spectrum after applying deep learning Weiner deconvolution to the chromatic confocal microscopy data is shown. The deconvolved spectrum exhibits two main peaks, accompanied by other peaks of

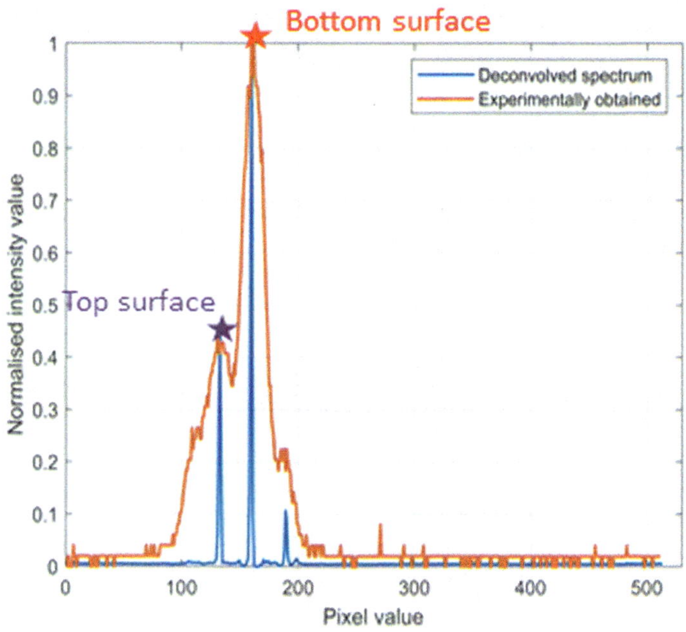

Fig. 11.14 Deconvolved spectrum for the grating sample

lower intensity. This phenomenon can be attributed to the sharp edges of the grating. When the light spot encounters the structure edge, it interacts with two different height positions (top–bottom), resulting in two wavelengths being detected by the camera sensor. Furthermore, the high NA of the objective influences the detection of additional wavelengths when the illumination spot moves from the bottom surface to the top surface (or vice versa), as more interactions between the outer cone of the light spot and the structure are introduced. The interplay between the high NA objective and the sharp edges of the grating plays a crucial role in shaping the spectrum's characteristics and poses challenges during the reconstruction process. Understanding these effects is vital for accurate and reliable measurements using deep learning Weiner deconvolution in chromatic confocal microscopy.

The intensity values of the peaks in the spectrum are influenced by the amount of light energy detected at each height level. When the light travels from bottom to top, more energy is concentrated on the bottom surface, with less energy reflected from the top surface. This results in two peaks in the spectrum, with the peak corresponding to the bottom surface having a higher intensity value, as shown in Fig. 11.14. During the reconstruction process, the peak position with the highest intensity value is considered the primary height.

Note that the above plot shows that the raw data obtained from the grating experiment exhibit a low signal-to-noise ratio (SNR) due to the limited light collected by the objective. However, by applying deep learning Weiner deconvolution, the resolution of the spectrum significantly improves, enabling easier detection of peak positions. This highlights the effectiveness of deep learning deconvolution in enhancing the quality and accuracy of the measured spectrum despite the challenges posed by the low SNR.

The three-dimensional plot obtained from the deconvolved spectrum is given in Fig. 11.15, with its cross section provided in Fig. 11.16. These visualizations demonstrate the enhanced quality and clarity achieved through the application of the proposed deep learning deconvolution technique.

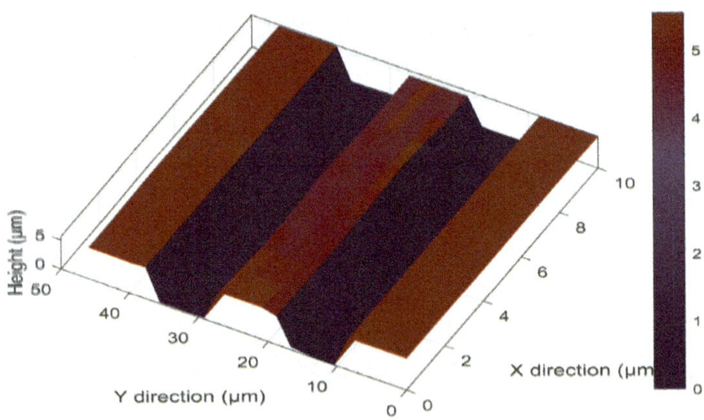

Fig. 11.15 Three-dimensional plot of the grating structure obtained from deconvolution

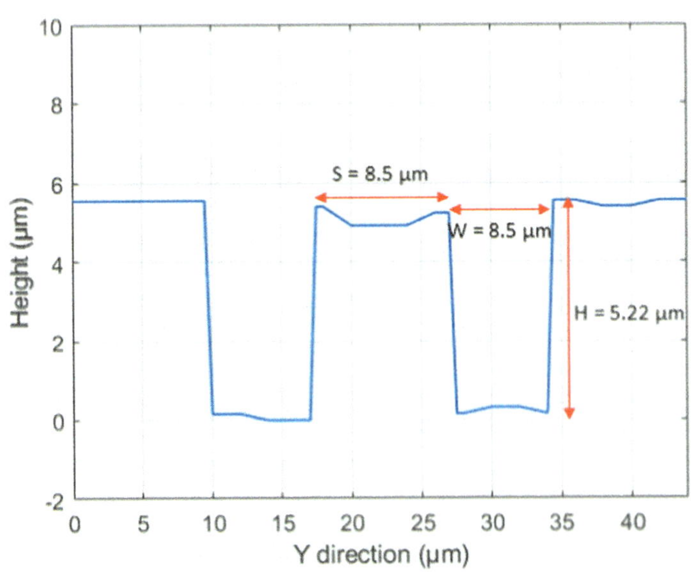

Fig. 11.16 Cross-section plot of the grating structure

11.5.4 *AFM for Standard Measurement of the Grating Structure*

AFM is renowned for its exceptional precision in sample measurement and thus was employed to perform reference measurements for the grating structure. The tip for the AFM measurements is shown in Fig. 11.17, having a type name of PPP-RT-NCHR-20. Detailed specifications of this probe are provided in Table 11.10 [3].

Throughout the measurement process, precise measurement conditions were established and configured based on the size and structure of the sample. Table 11.11 shows the assigned measurement conditions for the sample with a nominal dimension of W8/S8.

Figure 11.18 presents both the three-dimensional and two-dimensional plots of the measured sample obtained from AFM. These visualizations provide a comprehensive depiction of the sample's characteristics and structure. The cross-sectional plot of grating is shown in Fig. 11.19.

Table 11.12 compares the results obtained from AFM measurements and the deconvolution method, allowing for a comprehensive assessment of the performance and accuracy of the proposed method.

In Table 11.12, the observed bias between the data can be attributed to the significant influence of the moving stage during the measurement process. As the sample

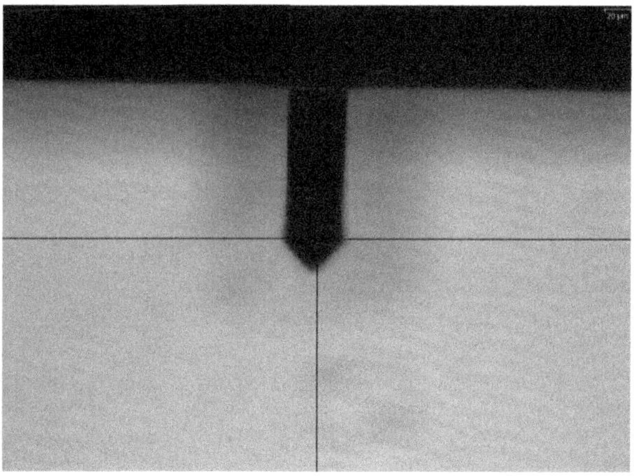

Fig. 11.17 Image taken for probe of AFM

Table 11.10 Technical specification of AFM probe

Parameter name	Value
Thickness (μm)	4 ± 1
Length (μm)	125 ± 10
Width (μm)	30 ± 7.5
Tip height (μm)	$10 - 15$

Table 11.11 Measurement conditions for measuring the W8/S8 sample with AFM

Name	Value
Data width (pixels)	256
Data height (pixels)	256
Fast scan axis	X axis
Fast scan direction	Left to right
Slow scan direction	Bottom to top
X scan size (μm)	32
Y scan size (μm)	32
Scan rate (Hz)	0.2
Z servo gain	2

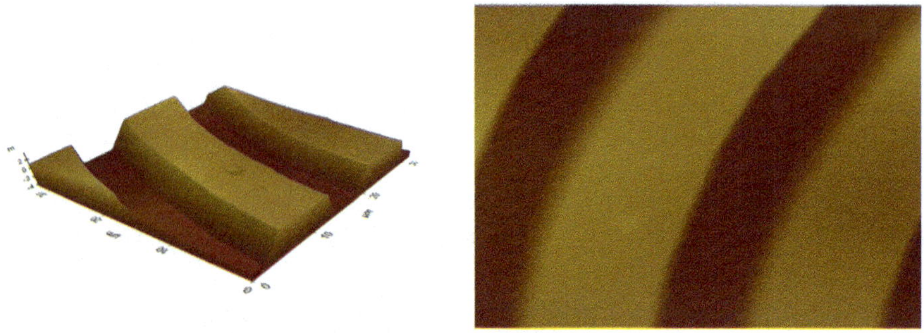

(a) (b)

Fig. 11.18 **a** Three-dimensional result of W8/S8 sample obtained from AFM; **b** two-dimensional result of W8/S8 sample obtained from AFM

is not perfectly static throughout the measurement, the dynamic nature of the stage introduces biases in the results. Additionally, the scanning step of the stage also contributes to these discrepancies.

11.5.5 Ball Grid Array (BGA) Sample

A ball grid array (BGA) bump sample was measured to test the capability of the proposed system for characterizing two-dimensional structures with sharp edges. Due to the FOV limitation of AFM, this sample could not be measured by AFM microscopy. Therefore, a laser confocal microscope (Keyence VK-X1100 model) was used to perform reference measurements for comparison with the results obtained by the proposed deconvolution technique, as shown in Fig. 11.20.

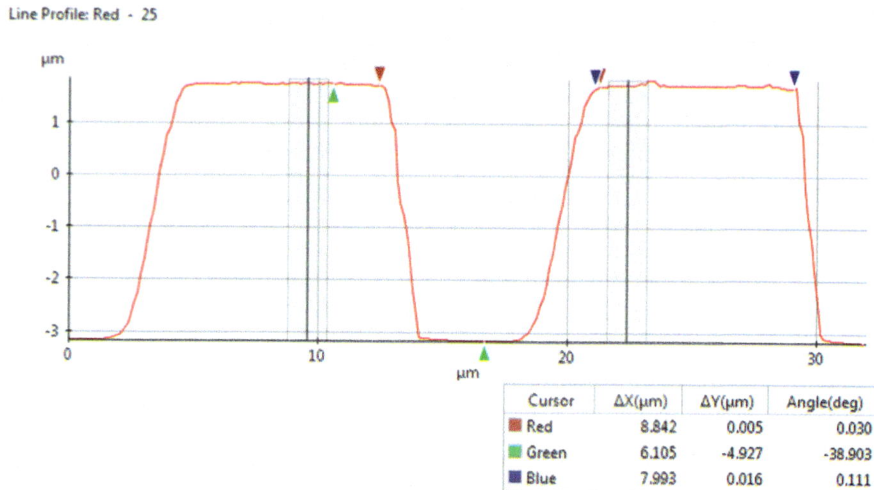

Fig. 11.19 Cross-section plot of W8/S8 grating sample obtained from AFM

Table 11.12 Results comparison between the deconvolved method and AFM

Dimension	AFM result	Deconvolution result	Bias
Width (μm)	7.993	8.5	0.51
Spacing (μm)	8.842	8.5	0.34
Depth (μm)	4.927	5.22	0.29

Fig. 11.20 Three-dimensional plot of the micro bumps obtained from Keyence

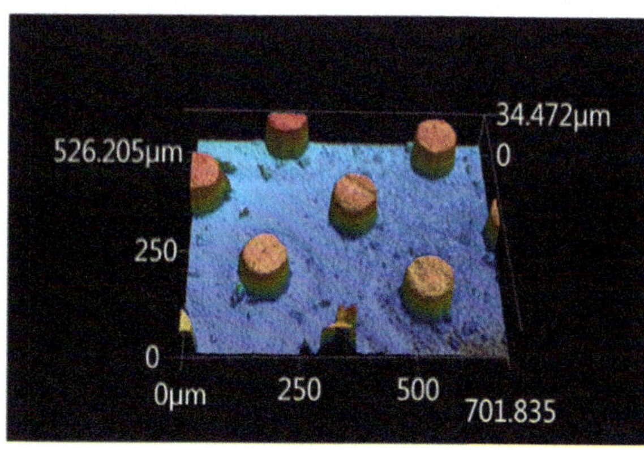

Due to its low reflectivity, reconstructing the BGA bump from the experimentally obtained spectrum is challenging. The spectrum obtained from the experiment is subjected to deconvolution, where the resultant spectrum is used to reconstruct

the three-dimensional profile of the BGA bump. Figure 11.21 shows the three-dimensional figure of the BGA bump obtained from deconvolution, with its cross sections along the *X* and *Y* directions shown in Fig. 11.22.

The discontinuity in the edges of the micro bump can be attributed to several factors inherent to the chromatic confocal microscopy technique. As chromatic confocal microscopy is a nonimaging technique, it can result in increased uncertainties when determining the exact edge locations from the analysis of the reflected light spectrum. The process of deconvolution, while effective in enhancing resolution, can also introduce some degree of blurriness from the deblurring operation. This blurriness can affect the sharpness of the edges in the deconvolved results.

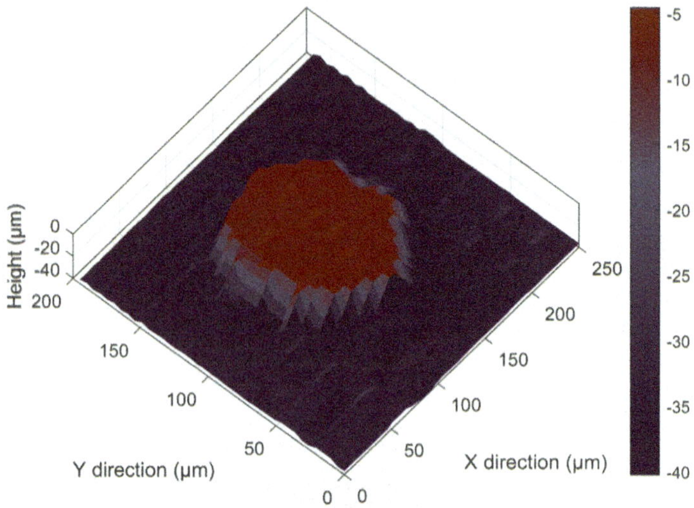

Fig. 11.21 Three-dimensional plot of the micro bump obtained from deconvolution

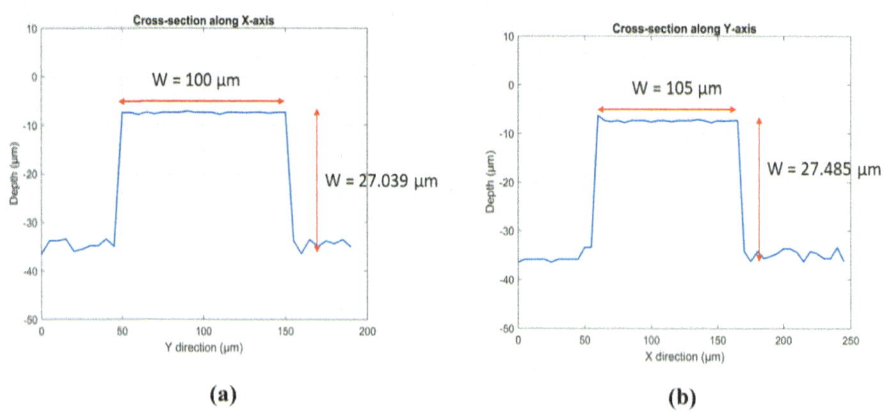

Fig. 11.22 Cross-section plot of the micro bump obtained from deconvolution

Additionally, the blurriness may also be influenced by the physical properties of the bump structures. Factors such as surface roughness, variations in reflectance, and nonuniformity in the sample surface all contribute to the perceived lack of sharpness in structure edges.

Despite the discontinuity in the edges, the deep learning deconvolution process still offers benefits in enhancing the accuracy of the measured depth information and improving the overall resolution of the bump structures. The deconvolved results can reveal finer details and features that might not be easily discernible in the raw spectrum.

In addition to the factors mentioned above, another crucial aspect that can influence the discontinuity in structure edges is the scanning step size of the stage during measurements. The choice of the step size plays a crucial role in determining the system's resolution and the level of detail captured during scanning. A larger step size can lead to a coarser representation of the sample surface, resulting in reduced resolution and potentially missing finer features of the bump structures. On the other hand, a smaller step size can improve resolution but may increase the measurement time significantly.

Consequently, to achieve the best possible resolution and minimize discontinuity, it is important to carefully optimize the scanning step size based on the specific characteristics of the sample and the desired level of detail. By finding the right balance between step size and measurement time, it becomes possible to obtain more accurate and finely resolved depth profiles of the bump structures.

11.6 Discussions

In this study, the accuracy and performance of the proposed deconvolution algorithm were examined. However, several factors may impact the measurement results of the system, as discussed in the following:

1. The deconvolution algorithm can be applied to curved surfaces, which are continuous objects. The dispersion lens must capture a wide range of incident light angles to obtain the necessary light information. However, the deconvolution effect on curved surfaces may not be as prominent as on distinct stepped surfaces, where the differences are more apparent.
2. When measuring rough surfaces, the reflected light received may exhibit variability, resulting in weak light intensity during deconvolution calculations. Strong light intensities can also easily influence the results. Increasing the threshold value can improve the calculation accuracy in such cases.
3. Measurements of objects with varying reflectance may result in poor image contrast and increased susceptibility to noise interference. To mitigate this, selecting a spectral detection range with better contrast can improve the system's adaptability.

4. Measuring structures with shallow depth and high aspect ratios presents challenges due to the limited light collected by the objective lens. The proposed deconvolution algorithm shows promise for accurate measurements in such cases. Considerations such as surface curvature, roughness, reflectance, and object characteristics should be considered to optimize the system's performance and adaptability.

References

1. Singh SK (2023) Theory and implementation of microscopic diffractive image profilometry. Master thesis, National Taiwan University
2. Chein W-H et al (2023) Non-integral depth measurement of high-aspect-ratio multi-layer microstructures using numerical-aperture shaped beams. Opt Lasers Eng 166:107563
3. Nanosensors (2024) PPP-RT-NCHR. https://www.nanosensors.com/pointprobe-plus-rotated-tip-non-contact-tapping-mode-high-resonance-frequency-reflex-coating-afm-tip-PPP-RT-NCHR. Accessed 2024

SPRINGER NATURE

GPSR Compliance

The European Union's (EU) General Product Safety Regulation (GPSR) is a set of rules that requires consumer products to be safe and our obligations to ensure this.

If you have any concerns about our products, you can contact us on ProductSafety@springernature.com

In case Publisher is established outside the EU, the EU authorized representative is:

Springer Nature Customer Service Center GmbH
Europaplatz 3
69115 Heidelberg, Germany

The manufacturer's authorised representative in the EU is Springer Nature Customer Service Centre GmbH, Europaplatz 3, 69115 Heidelberg, Germany. If you have any concerns regarding our products, please contact ProductSafety@springernature.com

Printed and bound by CPI Group (UK) Ltd, Croydon, CR0 4YY
30/09/2026
02225308-0020